CAREER MANAGEMENT FOR SCIENTISTS AND ENGINEERS

CAREER MANAGEMENT FOR SCIENTISTS AND ENGINEERS

JOHN K. BORCHARDT

AMERICAN CHEMICAL SOCIETY

OXFORD
UNIVERSITY PRESS
2000

OXFORD
UNIVERSITY PRESS

Oxford   New York
Athens   Auckland   Bangkok   Bogotá   Buenos Aires   Calcutta
Cape Town   Chennai   Dar es Salaam   Delhi   Florence   Hong Kong   Istanbul
Karachi   Kuala Lumpur   Madrid   Melbourne   Mexico City   Mumbai
Nairobi   Paris   São Paulo   Singapore   Taipei   Tokyo   Toronto   Warsaw

and associated companies in
Berlin   Ibadan

Copyright © 2000 by American Chemical Society

Developed and distributed in partnership by
the American Chemical Society and Oxford University Press

Published by Oxford University Press, Inc.
198 Madison Avenue, New York, New York 10016

Oxford is a registered trademark of Oxford University Press

Library of Congress Cataloging-in-Publication Data
Borchardt, John K., 1946–
Career management for scientists and
engineers / by John K.
Borchardt.
p.   cm.
Includes bibliographical references and index.
ISBN 0-8412-3525-2
1. Science—Vocational guidance.
2. Engineering—Vocational
guidance.   I. Title.
0180.55.V6B67   1999
502'.3'73—dc21   99-27070

9 8 7 6 5 4 3 2 1

Printed in the United States of America
on acid-free paper

To my parents,
my faults are my own,
my success I owe mostly to you.

PREFACE

The professional employment work of scientists and engineers has changed greatly in the 1990s and will continue to do so. It has become a truism that the era has ended in which scientists, engineers, laboratory technicians, and other technical professionals could expect to work for a single employer and retire at age 65 after a successful and productive career. Currently, society is learning to adapt to a new world of rapidly changing technology, sudden priority shifts, a renewed focus on the customer, more limited budgets, downsizing, and reengineering. Thus, technical professionals must reengineer their career goals and work habits, as well as develop professional survival skills. This book is intended to help in this effort.

Bemoaning the loss of the unattainable is not productive. Yes, many of us matured in a more orderly and predictable world. Many students still expect this extinct employment world to be awaiting them on graduation. However, what they will find is a new employment world. Supervisors (the ones that remain after downsizing) have much less time for mentoring and guiding their staff's careers as they manage increased numbers of people while looking for new ways to add value to the campany. To an increasing extent, we are all on our own. To say that we must be adaptable is a statement that tells us little. we must be more self-reliant and alert to new opportunities. Job security today is no longer the assurance of having a particular job until we decide to change jobs, it is the ability to find a new and rewarding job when we want to. We must be prepared to enter the job market more frequently than we had in the past. This translates into keeping our resumes updated and learning and utilizing current job-hunting techniques.

While I write from the perspective of a midcareer bench scientist, I try to include the concerns and needs of undergraduate and graduate school students as they complete their academic careers, as well as laboratory technicians and older professionals considering alternative careers that still involve science and engineering. I have tried to address the needs of engineers, B.S./M.S. chemists, and those who work in a variety of technical fields, as well as the concerns of those working in research and development (R&D) laboratories. Technicians are no longer limited to a career in the laboratory and are able to both work more independently at many companies and find an increased number of non-laboratory career options available to them as they gain experience.

This book begins by assuming you are technically competent in your field. Technical ability remains the foundation on which a successful career in science and engineering is built. Granting this assumption, I take you through the process of successfully managing your job and career. To have a prosperous and enjoyable career, you must begin by being productive and successful in your current job. Thus, much of what I discuss is actually job skills. However, these skills are useful at each stage in your career. Without mastering these skills, your career planning and management will be unproductive. This realization is the basis for organization of the book. The job skills you will meed throughout your career are discussed in Part I. Part II reviews today's work environment and evolving work culture. Job-hunting techniques are described in Part III. Only in Part IV is the actual process of career planning discussed. This is placed last in the book because mastering and practicing the job and job-hunting skills duscussed earlier is essential to a successful career and, thus, to career planning. However, some readers, especially experienced technical professionals, may wish to review the final chapter after reading chapter 1.

More specifically, the first chapter takes you through the changing structure of the employment world and how to set career goals. Critical tools for professional growth, such as time management and communications skills, are described in chapters 2 through 5. Cultivating your innovation skills and applying them productively is the subject of chapter 6. Skills needed to work productively with others are covered in chapter 7. Chapter 8 returns to the new employment world, in particular, the "horizontal" corporation. Team skills and how to work productively in the horizontal corporation are described in this chapter. Other skills for succeeding in the horizontal corporation's environment of empowered employees are discussed in chapter 9.

Chapters 10 and 11 cover the subjects of alternative career options, managing your own career, and maintaining your marketability should you decide to change jobs. Various aspects of job hunting are discussed in chapters 12 through 17. Like the employment world, many job-hunting practices have changed. While most old practices retain much of their value, new procedures such as storing your resume on disk with resume databases that employers may now search have changed many job-hunting techniques.

Technology career options are becoming ever more diverse. These options, deciding between them, and beginning an alternative career are described in the final chapter of this book. Experienced professionals uncertain about their career goals or students in the process of formulating their career goals may wish to read this chapter immediately after reading chapter 1.

Additional reading and other resources are presented at the end of most chapters. These are intended to be helpful, but not to be comprehensive. New books on career-related subjects are appearing at such a rapid rate that a comprehensive list would require a monthly update.

Acknowledgments

Even a scientist as great as Isaac Newton admitted that he did not succeed on his own, but saw further because he stood on the shoulder of giants. That is true for much more modestly successful scientists, as well. Many people play a critical role in our success. I would like to thank my parents who taught me how to care about others. The many chemists and engineers who served as mentors and role models indirectly contributing to this book are too numerous to list. However, a few deserve special mention.

Professor Robert Filler, at a dinner to celebrate my class' completion of our bachelor degrees, told us, "We haven't taught you much chemistry. What we've taught you is how to learn chemistry." My undergraduate research advisor, Professor Sheldon E. Cremer, was very patient with a novice researcher and guided my efforts in applying to graduate school. It was working with Shelly that my interest in chemistry turned into an enthusiasm that has flickered occasionally, but never burned out. My Ph.D. thesis advisor, Professor William H. Saunders, Jr., remains a role model of all that a scientist should be. He was patient with an often over enthusiastic graduate student when I tended to jump to premature conclusions. Both he and my postdoctoral research advisor, Professor Daniel J. Pasto, allowed a maturing scientist increasing independence and responsibility. Dan also was my first teacher in juggling multiple responsibilities as I watched him combine family activities, teaching, research, and textbook writing (as well as playing hockey) in a very active life.

One of my supervisors at Halliburton Services, Bill Young, taught me how to maintain enthusiasm and deal with career plateauing. Finally, one of my supervisors at Shell, Dr. Edward P. Rosenquist, has helped me overcome my naturally authoritarian work style and adapt to the new industrial chemistry R&D world of multifunctional teams and empowerment.

More directly, it is a pleasure to acknowledge tha advice and patience of Barbara Pralley, formerly Acting Manager, Book Acquisitions for ACS Books. Her enthusiasm for this project fueled and maintained my own eagerness. I also greatly appreciate the patience of my editor Charles Trowbridge of STM Publication Services.

Contents

CAREER MANAGEMENT FOR SCIENTISTS AND ENGINEERS

1

DEFINING YOUR CAREER

PURPOSE AND GOALS

Your career begins with setting your purpose and career goals. Your purpose is the end objective. It might be to develop a cure for cancer; it could be to make enough money that you can retire at age 40; it could be simply to fashion a career that you enjoy. Whether noble, sublime, or mundane, your purpose is set by your personal values. To enjoy a satisfying career, your purpose and goals must be consistent with your personal values. If your career path or its purpose is inconsistent with your personal values, this aspect of your life will seldom be satisfying.

Your goals are the rungs on the ladder you will climb to achieve your career purpose. To change the analogy, view your goals as intermediate signposts you pass while traveling through life striving to achieve your career purpose. It is an accomplishment to reach each of these goals (pass these signposts), but your trip is not over. There is always the next goal beckoning you. Well-chosen goals are indispensable in achieving your career purpose.

For example, suppose your purpose is to develop a cure for cancer. For most, to achieve this purpose means going to college, majoring in chemistry, biology, biochemistry, or a premedical curriculum. You may even see how an engineering or physics major can help you achieve your career purpose. So your goal at age 18 is to get into a good college. The next goal is to choose an appropriate college major. The goal after that is to achieve high grades to *perhaps* get you admitted to medical school.

However, your goals can change with time. For example, suppose you do not gain admission to medical school. You might still achieve your purpose by going to graduate school and studying chemistry, biochemistry, or microbiology. By working in the appropriate field, you can still prepare for a career in medical research. To return to our analogy, you can change paths and still travel toward the same destination—a career in medical research.

This example begins with the illustration of two points; your purpose and goals can change with time. A change in career goals is often based on the outcome of previous goals. It may also be the result of changes in your personal life, for instance, a chemist whom I respect very much changed his purpose and thus his goals relatively late in his career. When I knew him, he was well launched on a typical academic career and had just become a full professor. He stayed at the same university and later became chair of the Department of Chemistry and dean of the College of Arts and Sciences. He also supervised some graduate student researchers.

However, it was clear his primary goals were no longer focused on research. After losing his wife to cancer, his purpose changed. By returning to chemical research and synthesizing new families of compounds for medical testing, he hoped to spare other families the grief and tragedy his family had experienced.

In his best-selling book, *The 7 Habits of Highly Effective People,* Stephen Covey notes that you cannot separate goals and purposes. The two are interdependent. (In other words, the ends do not justify the means.) Your purpose must be consistent with your values and with the other purposes in your life. When this changes, or you discover values you did not know you had, you may redirect your career and change your goals. The case of the professor discussed above is an example. His personal tragedy led him to change his purpose from helping and guiding college-aged youth to finding a cure for leukemia.

Negative experiences such as a key college course failure, job loss, divorce, a serious personal illness, and so on, can also force you to reexamine your purpose and career goals. Positive experiences that lead to major life changes such as your children achieving adulthood and independence, "the empty nest," can also result in a reexamination of goals. The result of this reexamination may be a renewed determination to accomplish your professional purpose by changing your strategy and tactics. This may or may not involve changing your intermediate goals. For example, as governor of Arkansas, President Clinton lost his first reelection campaign. He did not change his goals or purpose to become president of the United States. However, a study of his career indicates he changed his tactics by becoming more willing to accept voters' priorities and not always stubbornly insisting that others accept his views.

Fortunately, many changes of purpose and career goals are the result of positive life experiences. For example, if you aspire to raise a family and be a good parent, the birth of a child can alter your career. You may postpone your career goals by staying at home to raise your child for an extended period of time. You may alter your goals by interrupting or ending your education and working full-time to support your family. Or, you many find creative ways to be both a parent and a chemical professional by working part-time, on a contract basis, or by job-sharing.

An important point is flexibility. While persistence is often a virtue, so is flexibility. The main issue at key points in your career concerns whether you (and your loved ones) will be happier if you remain determined to achieve the same goals or will you be happier by realistically altering your career purpose and goals. Answering this question has shaped many careers.

DEFINING YOUR PURPOSE AND GOALS

It is very difficult for others to help you set your career purpose. Often, it occurs by a process of osmosis as you observe different professionals and thus decide what you like and dislike about their careers. As you adopt role models, you often adopt their goals. (This may happen even if you remain unaware of their career purpose.) One reason for career dissatisfaction is that you may have well-defined goals, but a poorly defined purpose.

Goals are the steps by which you come closer to accomplishing your purpose. The process of defining and setting goals is discussed in chapter 7. Goals should be satisfying in and of themselves. Every goal need not propel you closer to achieving your purpose. Slight detours on your drive to your destination can be enjoyable. However, these goals should not substantially delay accomplishing your career purpose.

Goals should be ambitious, yet achievable. They should also contain a means of measure so you know when you have achieved them. Certainly, graduating and receiving a college degree covers both criteria. As is obtaining the type of job you set as your goal. Earning a certain amount of money, authoring a specific number of technical presentations, being awarded a particular number of patents, earning a promotion—can all be ambitious, measurable, and achievable goals.

Goals should provide pleasure and satisfaction, and the above list should satisfy most committed chemists, engineers, and technicians. If it does not, individuals need to reexamine their career goals and purpose. In the example cited above, the individual should perhaps consider teaching or beginning a business-oriented career in marketing and sales, or working in a production plant.

For most of us, our general purpose can be described as a rewarding career in the sciences or engineering. This achievement is usually difficult. The rewards are the professional achievements of scientists, engineers, and technicians—discoveries or technical accomplishments that better lives, provide jobs, and produce financial rewards to investors.

Unlike the 40 years after World War II, it is becoming more and more difficult to achieve a rewarding career. The workplace environment of most scientists, engineers, and technicians is rapidly changing. This is true whether you work in industry, government, or academe. It is becoming an environment in which professional and financial rewards are more difficult to achieve and more frequent career assessment and goal changing is necessary. Your purpose may change more often as well.

What are these environmental changes that are altering our careers?

CAREER PATHS
Academic Careers

Science and engineering careers traditionally have followed three major paths. The first is the academic career path most familiar to chemistry students. At the college and university level, research and education can dominate these careers. Teaching is more important than research at some smaller universities, four-year colleges, and two-year colleges.

When considering academic careers, do not ignore teaching at high school and grammar school levels. It is important to educate both the next generation of scientists and engineers and a technically aware citizenry who value the benefits of new scientific developments. This education takes place in our grammar and high schools. Leaders of the science and engineering professions recognize the critical importance of science treachers, many of them chemists, in these institutions.

Academic careers are changing also. For example, writing grant applications and shepherding them through the approval process takes an ever-increasing amount of faculty time at many universities and colleges. Young professors soon find that written communication skills are critical to the success of a grant proposal. As much research becomes more interdisciplinary, the ability to network and develop joint research programs with scientists and engineers in other fields is increasing the importance of oral communication and information science skills.

Funding issues are a major concern at many universities. Many universities advertise "anticipated" faculty openings for which the funding, and the actual position, never materializes. The growing use of adjunct professors (and the resulting decreased hiring of full-time assistant professors) has long been an issue in the humanities and may be increasing in science and engineering. Both funding limitations and employment conditions promote the hiring of relatively low-paid adjunct professors. This difficult job market increases the pool of

◈ Technicians and Teaching

Grammar and high school science teaching is not only for scientists and engineers. Some technicians spend an early portion of their career in secondary and primary school science education. An excellent laboratory technician whom I worked with, a biologist by training, provides an example of the benefits of teaching science early in one's career. A very hard worker (she is one of the few employees whose picture hangs in the Shell Oil Company museum), Judy Ardoin used to laugh at the rest of our research team when we would be stressed by the uncertainties of corporate reengineering. "Stress?" she would say, "You don't know what stress is until you've taught science to a bunch of teenagers!" Judy's teaching experience certainly helped her cope with the uncertainties of modern industrial research life.

young would-be academics as more freshly minted Ph.D.'s graduate. In addition, many industrial chemists are retiring at a young age and some go on to pursue academic careers.

The most junior of the stages of college and university academic career options, graduate school and postdoctoral research, are also changing. Due in part to diminishing career opportunities, these stages of a scientist's career are lasting longer than was the case even 10 years ago.

All these factors mean that even the most traditional career options, academic science and engineering, are changing.

Industrial Careers

The second major career path is that of industrial science and engineering. This career path offers many alternatives. There is industrial research and development. The development part of R&D is gaining more importance. Bench chemists, engineers, and technicians are becoming ever more involved in development activities. Beside conventional applications work, development can include customer technical service projects. R&D now often requires working on teams with people from other departments and other companies. As chemical professionals engage in this, the job skills they need are changing. For example, project teams now often include plant engineering and operating staff, marketing and sales personnel, government regulations and environmental specialists, and other professionals. As a result, the ability to present technical information to others who do not share that technical specialty is becoming much more important to career survival and success for all the team members.

The areas of government regulations, toxicology, and environmental chemistry are ever more important career opportunities for many chemists. Federal and state governments have passed much environmental legislation over the last 25 years. The resulting industry response has created additional employment opportunities for chemists, other scientists, engineers, and technicians. Many companies fund their own environmental chemistry programs to assure that their products and manufacturing processes have a minimal adverse effect on the environment. Various industries have also funded such research through trade associations. This research is performed at universities and research institutes, as well as industrial laboratories. Entire new industries such as site remediation and paper and plastics recycling are providing new employment opportunities for chemists, engineers, and technicians.

Traditionally, many industrial chemists, engineers, and technicians have worked in manufacturing plants in chemical, pharmaceutical, food, and other industries. They manage process equipment and provide quality control analyses to be sure products meet specific requirements. Other areas long open to these professionals include chemical sales and business management positions in marketing and other areas.

As everyone becomes more customer-oriented, chemical professionals are finding it increasingly important to speak customers' technical language and understand their technology. As companies reengineer, industrial chemists, engineers, and technicians are finding they have to reengineer their careers. This is discussed at length in chapters 8 through 11.

Trends associated with reengineering such as outsourcing and "spinning off" some operating divisions are providing new career opportunities for entrepreneurs and consultants. Extrapolating from the respondents to the ACS 1995 Salary and Employment Comprehensive Survey, Mary Funke, head of the Office of Career Services, found that 27,000 ACS members reported some income coming from consulting. This is about 18% of the total ACS membership. Clearly, consulting offers both primary and secondary incomes for many chemical professionals. This is further discussed later in this chapter.

Government Careers

The government has long been an important employer of scientists, engineers, and technicians. While employment is primarily at the federal level, these professionals also work in state and municipal government. For many years, chemists and technicians have worked in government laboratory, administrative, and investigatory functions to assure the safety of food and drugs. In the past 25 years, the growth of environmental legislation at the federal and state levels has led to the establishment of the U.S. Environmental Protection Agency (EPA) and many state EPA's. Many of these organizations have set up laboratories and staffed them with chemists, other scientists, and technicians to engage in the R&D necessary to establish environmental regulations and monitor their enforcement. (The effects of these regulations on industry career opportunities were previously discussed.)

Regulation of other industries such as mining also resulted in the government employing scientists, engineers, and technicians. With the energy crises of the 1970s, the government established the U.S. Department of Energy. While

---

◆ Management Opportunities

Many chemists, engineers, and technicians employed by the government work in laboratory positions. However, a considerable number of these professionals are administrators and managers. Like industry, many government agencies had a dual ladder for promotions and career development. As people advance, they have the option of remaining in the laboratory in a more senior position or becoming an administrator.

some of its laboratories had been in existence a long time, they were expanded and more chemists, engineers, and technicians were hired. There was also increased funding of university energy-related research.

Of course, the Cold War led to a vast expansion of government-funded defense R&D after World War II. The defense industry performed much of this R&D. However, government laboratories also performed a substantial portion. With the end of the Cold War and the need to reduce government deficits, both the defense industry and government laboratories are engaging in reengineering and downsizing similar to that experienced in the chemical and other industries. This will have a major effect on employment opportunities and career paths of chemical professionals who wish to work in government.

Variable Career Paths

The example of consulting illustrates the many career options open to qualified chemists. Nontraditional careers are mentioned many times in the pages that follow. Specific career options are discussed at length in the book *Career Transitions for Chemists*. (See the Additional Reading suggested at the end of this chapter.) The two keys to making a successful career transition are a realistic choice of another career and careful management of the transition process. As management structures are "delayered" and traditional advancement opportunities are reduced in industry and government (see chapter 6), there is increased interest in alternative career paths to professional fulfillment and financial security.

There are several transition processes. These can be summarized as follows:

- Concentric careers
- Contingency careers
- Combination careers
- Concurrent careers

Concentric careers are based on using core knowledge in two different activities. For example, a chemist might move from the research lab to the plant where he works on optimizing the types of chemical processes he studied in the laboratory. Later, the chemist can capitalize on his knowledge of plant operations to move into a production management position. Another example is the researcher who moves into applications research on the products he developed. This position could be technical service oriented with the chemist calling on customers and attending customers' trade association meetings. Continuing to capitalize on his existing knowledge and experience, the chemist eventually could move into a sales or management position in the same product area. Later, the chemist could apply the sales and management skills he developed to other product areas and more senior positions.

Contingency careers are careers that develop as alternatives should one lose their position or cannot obtain the type of job originally wanted. Personal motives often determine whether a career fits a contingency category. For example, a student who thinks her grades are too low to gain admission into graduate school or decides her lab skills are too weak for a career in research may take a minor. Examples include a minor in education to prepare for a science teaching career, a minor in business administration to prepare for a career in technical sales, or a minor in journalism to prepare for a career in science writing.

Working chemists may opt to go to night school to prepare for a contingency career. Being active in professional society activities can also be a good way to make the contacts you need to study contingency career options.

Many chemists have positions in which they combine features of two or more careers. For example, an applications chemist may supervise one or more technicians, often visit the plant to supervise the manufacture of product lots made to a particular customer's specifications, and join their sales representatives in calling on customers. These diverse responsibilities require diverse skills. These include personnel management skills, time-management skills, negotiation skills, and the ability to effectively respond to rapidly changing circumstances. (These circumstances could be as minor as making alternative travel arrangements midtrip when a flight is canceled to responding effectively when a prospective customer suddenly cancels a joint R&D program with your firm.)

Chemists who hold such jobs and develop diverse skills could be thought of as having combination careers. Developing the broad range of skills needed can position the chemist to change careers. In our example cited above, the chemist has positioned himself to transfer to sales, business management, or chemical production.

Some chemists have more than one career at a time. The second or concurrent career may or may not produce income. As noted earlier, many chemists who work full-time also consult part-time. Professional society activities are another example that can provide management experience and develop teamwork skills. Some view civic and other volunteer activities as a concurrent career even if they are not paid for their work. Consulting, or working on temporary and contract assignments can also be included in our list of concurrent careers. The institution of flexible hours at many firms is making it easier for many technical professionals to pursue concurrent careers.

One common feature of combination careers is a succession of professional temporary positions. Consultants and others who work under contract or in temporary situations often face the need to sell themselves well to obtain their next position. Good people skills are necessary to succeed in this situation. This is true in both job-hunting and on-the-job situations in which leaving a good impression, will help a chemist gain future assignments at the same firm.

A concurrent career is often a way for chemists to ease into a full-time alternative career without cutting off their prime source of income before they are ready. Part-time effort on a concurrent career can develop the skills and establish the network you need to succeed.

CHANGING CIRCUMSTANCES, CHANGING CAREERS

In today's career climate, changing your career path, and perhaps your career goals, is necessary much more often than in the past. Industry, large and small companies alike, is experiencing downsizing, mergers, divestitures, and other changes more frequently than years ago. This can result in very competent professionals finding themselves in the job market (see Figure 1). In the 1990s, the number of chemists who have spent part of a calendar year unemployed has steadily increased to 7.5% for calendar year 1995. (Data were taken from the American Chemical Society (ACS) 1996 Comprehensive Salary and Employment Survey.) If the 7.5% of the survey respondents is representative of the approximately 150,000 members of the ACS, this represents more than 11,000 chemists who experienced unemployment sometime during 1995. Career planning is needed so scientists and engineers have a defined plan of action should they find themselves unexpectedly in the job market. They need to keep their resume updated and their job-hunting skills sharp to get the type of job they want in a short amount of time (chapters 12 through 17).

In addition, corporate cultures at many firms are changing as they institute various programs to empower their employees while reorganizing to accomplish company goals through team, rather than individual, efforts. What may have

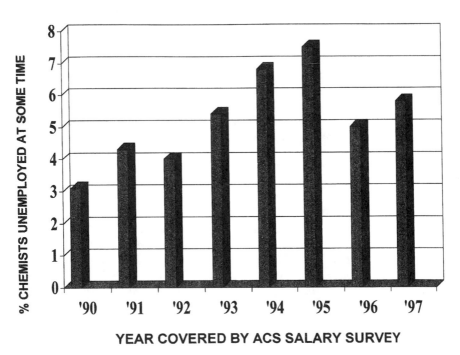

Figure 1. More chemists are experiencing periods of unemployment.

been a comfortable employment situation in the past may become uncomfortable in the future should you be unable or unwilling to adapt to these changes.

The result of these alterations in the work environment is more frequent changes in job responsibilities and employment. Also, job security now translates into the ability to obtain a professionally rewarding position in a reasonable amount of time should you lose your current position.

Important changes in your personal life: marriage, birth of a child, divorce, etc., are all events that can lead you to alter your professional goals and your career path. To adapt to such changing circumstances and achieve professional and career advancement, scientists and engineers have to use professional growth tools. These will be discussed in the next chapters.

ADDITIONAL READING

Covey, S. R. *The 7 Habits of Highly Effective People.* Simon & Schuster: New York, 1989.

Hall, D. T. *The Career is Dead—Long Live the Career: A Relational Approach to Careers.* Jossey Bass/Pfeiffer: San Francisco, CA, 1996.

Hall, L. A. *Becoming a Manager: Mastery of a New Identity.* Harvard Business School Press: Boston, MA, 1992.

Johnson, M. *Getting a Grip on Tomorrow: Your Guide to Survival and Success in the Changed World of Work.* Butterworth-Henineman: Newton, MA, 1996.

Kelley, R. E. *How to Be a Star at Work.* Times Business: New York, 1998.

McDermott, L. C. *Caught in the Middle.* Prentice-Hall Trade: Saddle River, NJ, 1992.

Moses, B. *Career Intelligence: The 12 New Rules for Work and Life Success.* Berrett-Koehler: San Francisco, CA, 1998.

Owens, F., Uhler, R., and Marasco, C. *Careers for Chemists: A World Outside the Lab.* American Chemical Society: Washington, DC, 1997.

Rader, C. P., Baldwin, S. P., Cornell, D. D., Sadler, G. D., and Stockel, R. F., eds. "Plastics, Rubber, and Paper Recycling: A Pragmatic Approach," ACS Symp. Series No. 609; American Chemical Society: Washington, DC, 1995. (Chapters in this book describe emerging technologies that could result in business opportunities for chemical entrepreneurs and employment opportunities for chemists, engineers, and technicians.)

Rodman, D., Bly, D. D., Owens, F., and Anderson, A-C. *Career Transitions for Chemists.* American Chemical Society: Washington, DC, 1995.

Thrailkill, D. *Temp by Choice.* Career Press: Hawthorne, NJ, 1996.

Udall, S. and Hilltrop, J-M. *The Accidental Manager: Surviving the Transition from Professional to Manager.* Prentice-Hall Trade: Saddle River, NJ, 1996.

Walesh, S. G. *Engineering Your Future: Launching a Successful Entry-Level Career in Today's Business Environment.* Prentice-Hall Trade: Saddle River, NJ, 1995.

Yeomans, W. N., *7 Survival Skills for a Reengineered World.* E. P. Dutton: New York, 1996.

PART I
PROFESSIONAL SKILLS

Successful science and engineering careers are based on the bedrock of sound technical knowledge continuously updated as necessary as well as performance in your current job. However, additional skills are necessary for a rewarding career. These skills are:

- personal productivity, particularly, effective time management
- communication skills (technical writing, oral presentations, and networking)
- innovation skills and decision making
- interpersonal skills

Personal productivity skills include effective time management, teamwork, and the ability to access needed technical and business information. These skills are discussed in chapter 2. Communication skills are covered in chapters 3 through 5. Innovation and decision-making skills are reviewed in chapter 6. Some interpersonal skills such as working on teams and resolving workplace disagreements are discussed in chapter 7. Subsequent chapters in Part II describe how to use these skills to achieve career satisfaction and professional success.

2

PERSONAL PRODUCTIVITY

Effective time management is only one of several personal productivity tools. Organizing your workplace and your files can also improve productivity, as well as working more effectively with coworkers (chapter 7). Communication skills can be critical in reporting results, convincing others to accept your point of view, and persuading prospective customers to evaluate the products or processes you develop.

EFFECTIVE TIME MANAGEMENT

Effective time management can add enjoyment to your job, reduce stress, and increase your productivity. This skill includes a collection of techniques, not all of which will work for you. The secret is to find and use those that will.

Effective time management begins with understanding how you use your time. To determine this, keep a time log for at least two weeks. Some professionals keep a continuous time log and review it monthly to see if they are developing poor time management habits. For your time log to be accurate, you must record the amount time spent on activities as soon as they are completed. Relying on your memory to complete a time log later will greatly reduce its accuracy.

When you have at least two weeks of data, analyze it and determine which activities occupy your workday or workweek and the amount of time spent on each. You need to make a concerted effort to change your work style. To do this, determine which activities do not contribute to your job performance, contribute to your profession, or improve your sense of personal satisfaction and well-being; stop engaging in these activities. Then determine which remaining activities can be delegated to others over whom you have some authority. (You may have to enforce this delegation.) Any duties assigned to others must be within their competence and appropriate for them to perform. By doing this, you will have more time for the higher-priority and more-challenging tasks that remain on your list.

Organize Your Time

Begin by keeping a list of projects and deadlines. Then, establish priorities. Modify this list as you complete assignments and receive new ones. Plan your work schedule around these priorities. Priorities should be based on project importance, its benefit to you and your employer, and the project deadline.

Use a "month-at-a-glance" calendar so you do not miss meetings or deadlines. This type of calendar also helps to avoid conflicts in schedules. When entering events, use a color code: red for deadlines, green for meetings, etc. Use this monthly calendar as the basis for a weekly plan and daily "to-do" lists.

To be effective, you will have to reserve time in fairly large chunks, two hours or more, for major tasks. Working on tasks intermittently reduces your productivity and the total amount of time required will be greater. Establish these blocks of time as regular features on your calendar. For example, you might reserve three hours every Tuesday, Wednesday, and Thursday morning for major assignments. Determine the time of day when you are most productive and reserve this time for the most lengthy, difficult, and important tasks.

Schedule tasks to fit the available time. Use short periods for telephoning, reading, and writing business correspondence. Dividing large assignments into smaller pieces will make it easier to use short blocks of time. With resource materials readily available, you can effectively utilize unexpected openings in your schedule.

Minimize self-created interruptions. Do this by grouping similar tasks together. For example, if you have several phone calls to make or business letters to write, do them all at once. This prevents short tasks from repeatedly interrupting longer ones and thus reducing your productivity.

When you schedule a meeting, set a clear ending time as well as a starting time. When someone schedules a meeting you need to attend, determine when this meeting will end and schedule your time accordingly.

Shut off distractions. Inform your coworkers that you do not want to be disturbed at certain times and these times are off-limits, except for emergencies. These should be your large blocks of time that you reserve for major tasks. Use your answering machine to screen telephone calls. Use these "off-limit" times for your most challenging assignments. Delegate appropriate tasks. Delegating less-technical tasks to a technician, librarian, or clerk can add variety and interest to their jobs, while helping you stay focused on your most-challenging assignments.

Learn to say no. Do not agree to deadlines you cannot meet or projects you have no time to complete. At either the end or beginning of your workweek, review the past week—what was accomplished and what was not. Check your priorities and "to-do" list. Then, put together a schedule for the upcoming week.

Handling Documents

Handle documents only once. When you read a document, determine the appropriate action. Take this action immediately and move on to your next task.

For example, respond to correspondence by writing an answer on the back of the letter immediately. Then, when you type a polished response later the same day, you will not have to reread the letter to focus your writing properly.

Store active files in a convenient location near your telephone. Place items for an upcoming meeting near your office door. This way, you can grab the items quickly on your way to the meeting.

Organize your files in a logical way. Do not create a separate file for each type of document. A researcher could organize project files alphabetically or by project number. Applications specialists and sales personnel could set up files alphabetized by customer name. Organizing separate files for each product might work best for people in marketing, sales, and environmental, health, and safety affairs. Color coding your files could also be helpful.

Before filing a document, ask yourself, "Where will I probably look for this when I need it?" If there is more than one good answer to this question, make copies of the document and put each in an appropriate file.

Keep a list of your most frequently called telephone and fax numbers. Post this in a visible place. Should you travel, bring a second list with you; perhaps storing it in a notebook or hand-held computer. Organize your office equipment and supplies with the most-used items closest to you and less-used items farther away.

Business Reading

Be realistic about how much you can read. This will help avoid stacks of unread technical journals, trade magazines, and business magazines building up in your office. Use a weekly or semiweekly on-line literature service to keep you updated on published information in your field. Use a service that provides article abstracts. Read the abstracts before deciding whether you need to read the full article. Knowing where to find full-text journals and magazines on-line can also help you limit the stacks of unread periodicals in your office.

Clip or make copies of important articles when you read them and file them appropriately. Clip or photocopy articles you want to read, but do not have time for at the moment. Use a simple priority system such as numbering articles "1, 2, and 3" to prioritize them. Place these in a visible location. Take some with you when you go to a meeting in case it starts late. You can also take this file with you to read during business travel, while waiting in a customer's reception room, while waiting in your doctor's or dentist's office, etc. Be alert to opportunities to catch up on your reading, even if time is short.

Use Computer Technology

A microcomputer and spreadsheet program can be an excellent time-saving tool for recording large volumes of data and performing calculations. Word processing is more efficient than writing by hand if you have to make many revisions.

Reorganize your computer files so major directories correspond to projects and other activities. Consider major directories to be file cabinets, secondary directories as file drawers, and tertiary directories and individual files as file folders. Use the same system on diskettes and other memory back-up systems. Always backup your files; the time you spend doing this can be repaid tenfold should your computer system crash. Clearly label back-up diskettes.

When using software, program "macros" for frequently used functions that do not have keyboard commands. For example use keyboard commands such as "Ctrl-I" for italics rather than entering multiple menu commands to accomplish the same task. Set up document templates in your word processor. These can be for forms such as invention disclosures and frequently typed documents such as technical service reports and cover letters, internal memos, etc. Keep a word processing file for ideas and reminders to yourself. Use this file when setting up daily to-do lists.

While particularly useful for sales personnel, a personal information manager can help all professionals organize and manage project and customer information. Using a personal information manager, you can store and retrieve all contacts with customers. These include summaries of telephone and face-to-face conversations, written reports, progress on technical service projects, and personal information about the customer's family and hobbies. Review this information before calling or visiting the customer.

If you download a large amount of information, purchasing a high-speed modem is a worthwhile expense. Should you replace a slower modem, make sure you properly set up your software to take advantage of the increased modem speed.

Use Communications Technology

Voice mail is, of course, an essential tool for professionals that allows them to leave the office and not miss important messages. However, proper voice mail

---

◆ Storing Personal Information

Is it manipulative to record personal information about a customer and review it before contacting the customer later? I do not think so. Even some people in long-term relationships would forget important information such as anniversaries and birthdays if they did not sometimes use notes as reminders. Taking the trouble to make and use personal notes shows you care about individuals just as you care about business problems when recording the technical information customers provide.

Personal relationships facilitate business interactions. Some of my customers have become friends—my partners in mutually rewarding business, professional, and personal relationships. The initial effort we made to get acquainted facilitated establishing these partnerships. Recording personal information promotes this "getting to know each other" process.

technique is necessary to avoid playing telephone tag and thus wasting time. Voice mail offers an excellent means of staying in touch with customers even outside normal business hours. When leaving a voice mail message, mentally rehearse the message before you call to be sure it will be clear and concise.

When returning calls that require a brief answer or making calls to ask a simple question, call during your recipients' lunch hour when many people are away from their desks. By leaving a message on their voice mail, you can limit the time you spend on the call. This is especially useful with a talkative respondent.

When you call with a complicated question and reach someone's voice mail, ask your entire question clearly and as concisely as possible. Include all specifics so the other party can provide you with a complete answer in a single telephone call.

When you must speak personally with respondents, leave a voice mail message asking them when they will be available to take your call. Identify the purpose of your call, this way they can prepare for your follow-up call. If the call is urgent, be sure to say so. After they leave you a message, call them at the earliest appropriate time. Also include in your message the times you will be available in the next 1–2 days. Then, be sure to be in the office at these times.

Subscribing to caller ID saves you from asking customers for their telephone numbers. Caller ID can also let you know if you should interrupt an important task to answer the telephone immediately or allow the call to go into your voice mail.

Electronic mail (e-mail) has long been a staple of in-house corporate communications. On-line services and the World Wide Web have made this technology widely available. With this technology, people can rapidly send information and data that is too lengthy or complicated to detail over the telephone. Many professionals have e-mail addresses and include them on their business cards. It is an inconvenience (and suggests that technologically you may be out of date) not to be able to send and receive messages by e-mail. An alternative to e-mail is the fax machine.

Use Travel Time Effectively

Chemists, engineers, and technicians are working more on teams with plant and sales personnel and spending more time working with customers. This means increased business travel. Increased travel time correlats to a need to become more productive while on the road. When you schedule a trip to a certain area, consult with all your customers in that area. Determine with which customers it would be productive to meet. This way, you can visit an area and meet with a customer during the same trip, thus avoiding a second trip later. Have a clear purpose in mind for each meeting so no one will feel you wasted their time when you leave their office.

Try to group appointments by scheduling them so you do not have to dash from one to another, but also so you do not build a lot of "dead time" into your trips. Consult with customers if you are unfamiliar with local geography and travel times.

Business reading and short writing projects can make time spent in airports, on planes, in hotel rooms and customer waiting rooms productive. Bring what you will need to work on at least two short projects. Then, if you complete one project or reach a dead end, you can still use your time productively. Notebook computers make it possible to do data analysis, extensive calculations, and word processing when traveling. With the appropriate software, you can have names, telephone numbers, and the customer information needed on your trip. You can also set up future appointments, call customers, answer voice mail messages, and do other work. This can be a lifesaver when someone calls during your trip and does not leave their telephone number on your voice mail. With the right software, you can even work on your expense account. With a modem, you can submit expenses before you return to the office. A modem will also allow you to read and answer e-mail.

At some companies, the professional's office computer is a notebook computer equipped with a docking station. So taking it on a trip simply means detaching it from the docking station. At some firms, travelers can check out a notebook computer just for use on business trips. If your company operates this way, be sure you travel with the appropriate software needed to do your work. If you have personal databases such as telephone lists on diskettes, bring them on your trip. (Be sure to leave a back-up copy at home.)

Prepare thank-you and follow-up letters and envelopes in advance of your trip and store them on your notebook computer. Edit them as necessary following your meeting. For tailored letters, typing the address and salutation before leaving on your trip can help get you started writing while on the road. You may wish to pack stamps, company stationery, and envelopes with you. This way you can prepare letters in your hotel room the evening after a meeting. You can bring a portable printer with you, use a printer that is available in some "business class" hotel rooms, or have your letters printed in a hotel business center. Alternatives include faxing letters to your customers or sending them by e-mail.

Instructional tapes and books on tape also offer a means of productively using your waiting and travel time.

Rewards

Reward yourself and express appreciation to others after completing a major project. Take your assistant or secretary to lunch to thank them for their help. If you did a large amount of project work at home, show your appreciation to your family for their cooperation in providing you the quiet time you needed to work. Take them out to dinner or a movie to celebrate the completion of your project. Expressing appreciation to coworkers and family members will make both you and them feel good. You will also get more cooperation when your next major project comes along.

YOUR PERSONAL LIFE

Saving time in your personal life also can reduce stress and give you more time for important family and career activities. When you have an appointment with a doctor, dentist, lawyer, or other professional, ask for the first appointment of the day. It is the one least likely to be delayed. Use the same strategy for non-emergency household appointments such as plumbing repairs.

Group your errands by location and combine trips. For instance, if you have grocery shopping to do, consider what other items you need and could buy at nearby stores. Keep lists, perhaps on your refrigerator, of items you need from the grocery store, hardware store, etc. Take your lists with you on shopping expeditions.

Purchase a selection of cards for appropriate occasions such as birthdays, anniversaries, weddings, and loss of a loved one. That way, when you hear about an important event in a friend's life, you will be able to respond immediately to show your concern.

Clearly separate your personal life and your business life. Have friends with different interests who do not necessarily work in your field or at your company. This does not mean to avoid socializing with coworkers off premises. It does mean that you will avoid "shop talk" in many social situations. This can reduce stress and make your life more interesting.

Similarly, separate your personal and business space. If you bring work home, set aside a specific location to work on it. This separation will help reduce stress. In particular, do not bring work into your bedroom. Doing so can make it more difficult to relax and sleep. Minimize bringing your personal life into your job. Although employers often allow for certain activities such as selling cookies to help your children raise money for scouting or other activities, avoid overdoing such activities and taking too much time away from your job.

Communications technology can also be a big timesaver for callers. Some ideas that consultants can use are described in the Communications Technology section above. In addition, software will permit your modem to automatically and periodically dial your on-line service and poll your e-mail address for messages. You can transfer these to your computer and store them to review at your convenience.

The broadcast feature in many fax machines can be an efficient way of sending multiple copies of a document. These can include updated rate sheets, announcements of times you will be on vacation and unavailable, and announce-

---

◆ Additional Tips for Consultants

Consultants can benefit from many of the tips listed in the section entitled "Your Personal Life." In addition, they should try to schedule meetings in client's offices if they are nearby. It is easier to excuse yourself and leave someone else's office when a meeting is no longer productive than to get visitors out of your office.

If you do schedule meetings in your office, remember that meetings should have scheduled conclusion times, as well as starting times.

Label your supply storage shelves so you will be able to find things easily and quickly, while determining if you are running out of a needed item. When you realize you need an item or have to run an errand, add it to your "to-do" list. Running errands is one of the few times where you should consider grouping business and personal tasks together. For example, when going grocery shopping, you might visit the nearby office supply store to pick up needed items. Comparison shopping at two or three stores to save a few dollars is not a productive use of your time. When you purchase business items, buy enough to last one to three months.

Take advantage of your self-employed status; shop and bank during off-peak hours.

Shopping by mail can reduce the time you spend running business shopping errands. If your budget can tolerate the expense, use delivery services to drop off business items, pick up overnight delivery envelopes, and run other errands.

Update your voice mail message to inform callers what time you plan to be back in the office when you leave to run errands or meet with a client.

To save time, rely on travel agents in scheduling flights, buying airline tickets, making hotel reservations, and arranging rental car reservations. Whenever possible, try to schedule business trips enough in advance to obtain a cheaper fare. For trips you cannot bill to clients, such as attending conferences, use your frequent flier mileage to obtain free air tickets and limit your business expenses.

There are many ways to take advantage of personal computer technology. These have already been described above. However, some deserve amplification for the individual consultant. Setting up document templates in your word processor can be particularly useful. These templates can be for forms such as customer bills and time records and frequently typed documents such as customer and cover letters. With care and a letter-quality printer, these forms can have a thoroughly professional appearance. By using different fonts and purchasing appropriately sized paper, you can eliminate the need to purchase business forms. Alternatively, you can prepare the format to print information neatly in the appropriate locations on a purchased business form such as an expense statement. With careful preparation of a letterhead and an excellent laser printer, you can eliminate the need to custom order printed paper and envelopes.

ments that you will be attending a particular conference. While social rules may be changing, communications that have a more personal or social nature such as the announcement of major accomplishments or changes of employment are best sent by mail or communicated by telephone call. Computer e-mail is becoming accepted for these career-related, but more personal, communications.

ACTIVE LISTENING SKILLS

Active listening skills can be considered either a communications skill or an interpersonal skill. However, because so much productivity loss occurs due to poor interpersonal communication, active listening is discussed in this chapter. *This bears repeating: Poor communication of information and instructions can result in significantly reduced productivity. Hard work is of little use if you are not working on the right problem.* A person's productivity can be significantly reduced because of a misunderstanding of how and what is supposed to be done. Employee empowerment has made effective listening skills more important as more instructions tend to be given orally in team meetings than as written documents. Downsizing has contributed to this as managers have less time to prepare written instructions for their staff members. Similarly, chemists and engineers have less time to prepare written instructions for the technicians they work with daily.

Effective Listening Skills

Using effective listening techniques, you can walk away from discussions and meetings with a clearer understanding of the information conveyed and what your responsibilities include. Effective listening can reduce misunderstandings and enable you to better focus your work. As more work is performed though teams, effective communication becomes even more important.

Effective listening starts before a meeting begins. Begin by entering each formal meeting (as opposed to impromptu hallway discussions) with a clear idea of both the purpose of the meeting and your objective in attending. This will help you listen more effectively and prepare more relevant questions. Apply this maxim to technical conferences, seminars, internal corporate meetings, and one-on-one discussions with your supervisor and coworkers. Prior to attending these types of meetings, I find it helpful to write down my purpose and any important questions I want to have answered. During the course of the meeting, do your best to have each of your questions answered. You may not succeed in this, but it is worth the effort. The answers to your questions will increase the value of the meeting to you and perhaps to other participants, as well.

Verifying Understanding

Being an active listener involves both verbal and nonverbal communication. Nonverbal communication includes an active posture, an alert, but not fidgety manner, and a pleasant (if appropriate) facial expression. These signal to the speaker and any others present that you are open-minded and want to hear what they have to say. (A pleasant expression may not be appropriate in some situations. Examples include discussions of why deadlines were not met, certain phases of performance discussions where bad news must be conveyed, and other situations where unwelcome information must be imparted to other meeting attendees.)

Nonverbal cues such as nodding or shaking your head and a quizzical expression let the speaker know whether he is getting through to you. This is important. Do not try to hide confusion or lack of understanding. How many times have you experienced someone who was speaking interrupt themselves to say, "You look puzzled," during the course of a conversation or presentation? This presents you with a golden opportunity to ask a question to resolve your confusion. Good teachers rely heavily on these nonverbal cues from their students.

Some brief verbal cues serve a similar purpose in conversations. You can provide these low volume verbal cues during pauses in the conversation. Examples of cues that let the speaker know you understand her point include: "Oh!," "Aha!," and "I see." Verbal cues indicating puzzlement, but may seem rude, include: "Oh?," "Huh?," and "What?". It is better to respond with a brief comment during a pause in the conversation. This may be a short question or simply, "I don't understand." Brief restatements can indicate you understand the speaker's remarks. As the inclusion of "Oh" in both lists indicates, voice tone and intonation are important in providing verbal cues. Of course, verbal cues are inappropriate and even disruptive when you are a member of an audience.

When you are speaking, watch your audience for both nonverbal and verbal cues. This is easy in a personal conversation, but may be impossible in a darkened room showing 35-mm slides. Respond to these cues. In a conversation or informal presentations to a small audience, you can ask the person providing the cue to state their point of nonunderstanding or confusion. Do so in a diplomatic, nonconfrontational way. Even a pleasantly delivered, "What don't you understand?" remark might seem confrontational or disrespectful depending on the person you are addressing. More effective is the mild, self-deprecating remark, "By the expression on your face, I haven't explained this well enough." Follow this by asking, "How can I explain this better?" or "What is it you don't understand?" The prefacing remark robs this second question of its sting. Voice tone is important in asking this question. Avoid sounding petulant or impatient and giving the impression you think the other person is stupid. This is also important when answering listeners' questions.

When you are dealing with a large audience and can see nonverbal cues such as facial expressions or restlessness, consider changing your presentation in response. This may mean simplifying the remaining portion of your presentation or (if you are not pressured by time) going back and reexplaining important points in different, perhaps simpler, terms.

When you are listening to someone else provide information or give instructions, restatement can be an effective technique to verify to yourself that you have correctly understood something. In this technique, you take the speaker's last remark and state it in your own words. This is often, but not always, in the form of a question. For example, suppose your supervisor tells you that before beginning a project you need to talk to Jim, but you are not completely sure about what exactly. Use the restatement: "What you are saying is, before I start work, I need to talk to Jim to find out what impurity levels our customer can tolerate." Your supervisor will verify your understanding if you are correct or restate himself more clearly if you have not understood him.

If the subject is complex, restatement of a single concept or instruction may not be sufficient to be sure you understand everything. In this case, use the technique of summarizing. Using this technique, you summarize a series of related remarks in your own words so the speaker can verify that you understood him correctly. For example, you could summarize your supervisor's remarks as follows: "What you mean is that if we proceed simultaneously with FDA approval for human testing and manufacturing product scale-up, we will be able to save six months. This means that I have to order the process units in time to have the pilot plant ready for test runs May 1." Summaries can also be in the form of a question to be better assured of getting a clear response. Adding "Right?" to the previous example forces the supervisor to respond. This response will either verify your understanding or correct it. Either way, both you and your supervisor will leave the meeting knowing that you understand clearly what you need to do and will work productively.

If you are the person providing information or instructions in a one-on-one discussion or small work group meeting, asking for a restatement or summary helps you determine if the other person understood what you said. This is best done after each important instruction or piece of information. You should also solicit a summary at the end of the conversation. Be diplomatic and avoid sounding confrontational. For example, saying, "Repeat what I said," sounds belittling and disrespectful. Instead conclude your remarks or instructions by saying, "I hope this is clear. Based on our conversation, what do you need to do?" Voice tone is important in soliciting this feedback. Avoid sounding impatient or critical.

Questions

Asking questions is the time-honored method of soliciting additional information and verifying understanding. Questions can be open- or closed-ended. Open-ended questions require an extended response and yield a lot of information. These questions often begin with "How," "Why," or "What." Close-ended questions usually result in a brief answer and may not result in much information. However, close-ended questions can be valuable in certain circumstances. Closed-ended questions often begin with "Who" or "Which."

Questioning should begin with open-ended questions to settle major issues and define options. For example, consider the open-ended question, "How can we deal with the high brittleness of this engineering plastic?" Another example, "How can you improve your job performance?" The response to such questions often provides several options on how to proceed to solve a problem.

Follow-up questions may be mixtures of open- and closed-ended questions. Suppose our polymer chemist has responded to the question given above and provided several options to reduce the brittleness of the engineering plastic. Her supervisor can ask the open-ended question, "What are the advantages and disadvantages of each option?" This can be followed by the open-ended question, "What are the relative costs of each option?" The supervisor can then ask the closed-ended question, "Which option is most likely to succeed?" and the open-ended follow-up question, "Why?" Using questions such as these to elicit the relevant information, the supervisor and polymer chemist soon share the information they need to determine priorities in studying the various options and solving the problem.

Open-ended questions tend to work best early in discussions when the basic parameters of a problem or situation are being determined. Closed-ended questions work best to elicit very explicit information and as a check to be sure information previously given is correctly understood.

Most supervisors value staff members who use the techniques of restatement, summarizing, and questioning because they allow the supervisor to be confident that staff members leave the meeting with a clear understanding of what they are to do. This understanding means improve productivity.

Other Effective Listening Skills

Other effective listening skills can also improve the understanding of information others provide in presentations and conversations. These include the following strategies:

Do not tune out dry subjects. Instead ask yourself, "What's in it for me?" The answer to this question can help you pay attention. If you cannot answer the question yourself, ask the speaker a polite version of the above question.

Do not tune out speakers because of their poor delivery. Force yourself to pay attention to the content of the presentation. A good way to do this is to take notes.

Do not be argumentative. Let a speaker complete the point he is making before disagreeing. Waiting will give you time to develop a polite, constructive comment or question rather than just sounding confrontational. Argumentative questioners make a poor impression on all meeting attendees, not just the speaker.

Keep an open mind. Try not to overreact to a speaker's emotional words. Instead, consider why the speaker is becoming this way. Consider what is said, not who says it. This can open your mind to useful information from unexpected sources.

Listen for ideas, not just facts. Sometimes the facts alone are secondary and the ideas the speaker develops from the facts are what is most important. This is often the case when the speaker uses a set of facts to develop options for future work or recommend a specific course of action.

Capitalize on the fact that thought is four times faster than speech. Continue to listen, but also mentally evaluate, summarize, challenge, and anticipate the speaker's remarks. Consider what the speaker's tone and choice of words mean. Prepare questions to ask.

Supplementing Active Listening Skills

Concluding a meeting by restating action items and who is responsible for them, as well as preparing meeting minutes that include these items is a good technique. It is also good to have the meeting secretary review the action items at the conclusion of the meeting. This way you can be sure the distributed meeting minutes will contain correct information. The same techniques should be used for decisions that are reached at meetings. Meeting minutes are a very useful tool for reporting these decisions to concerned parties not present at the meeting and explaining why and how decisions were made.

ADDITIONAL READING

Time Management

Axelrod, A., Holtje, J., and Holtje, J. *201 Ways to Manage Your Time Better.* McGraw-Hill: New York, 1997.

Bocher, D. D. *Get a Life Without Sacrificing Your Career: How to Make More Time for What's Really Important.* McGraw-Hill: New York, 1996.

Covey, S. R., Merrill, A. R., and Merrill, R. R. *First Things First.* Simon & Schuster: New York, 1994.

Douglass, M. E. and Douglass, D. N. *Manage Your Time, Your Work, Yourself.* American Management Association: New York, 1993.

Wilson, S. *The Organized Executive.* Warner Books: New York, 1994.

Effective Listening

Bechler, C. and Weaver, R. L. *Listening to Win: A Manager's Guide to Effective Listening.* Master Media: Edina, MN, 1994.

Bone, D. *The Business of Listening: A Practical Guide to Effective Listening.* Crisp: Menlo Park, CA, 1995.

Murphey, K. *Effective Listening: Hearing What People Say and Making It Work for You.* Simon & Schuster: New York, 1988. Audiotape.

3

Writing Skills

As chemists, engineers, and technicians work more in teams and with their customers, they are communicating with more people from different professional backgrounds and with different degrees of technical expertise than in the past. Increasing workforce diversity also places increased demands on communications skills. The "delayering" of management means that staff chemists and engineers more often have to "sell" project ideas or action items resulting from successful projects to high level managers. The intermediate-level managers who used to perform this task are no longer at many firms. As a result of these trends, both written and oral communications skills are becoming important in the new work environment. Networking is, in part, a communications skill and, in part, an organizational skill. The workplace trends of professionals making more frequent job changes and increased outsourcing have made networking even more important for career success. Three communications skills: writing, oral presentations, and networking, will be discussed in separate chapters.

TECHNICAL AND BUSINESS WRITING

Numerous books have been written on technical and business writing (see Additional Reading at the end of this chapter). The key principle of effective technical and business writing is to understand who your readers are and write to satisfy their informational needs. Your goal must be not only to be understood, but also to write clearly so that you cannot be misunderstood. Business managers who try to decide whether to commercialize a new chemical manufacturing process and chemists who read the *Journal of the American Chemical Society* may both be reading about the same piece of research, however, the different informational needs of these two groups signify that the two different documents they read will bear little resemblance to each other in content, level of detail, or organization.

Effective writing consists of the following steps:

- Define the message or information to be communicated
- Determine the audience for your message
- Prepare an outline for your document. This usually is not necessary for short letters and e-mail notes. If your document is an article for publica-

tion, your outline should take into account the manuscript requirements of the journal or trade magazine.
- Write the document using the appropriate style and follow the organization of your outline.
- Revise your manuscript. Do this critically. Delete sections that do not contribute to the informational needs of your readers. For example, business managers and plant engineers do not need to understand the supercritical fluid chromatography techniques used to determine the level of byproduct X present in a pharmaceutical intermediate. What they need to know is the answer to the question, "Is the level of byproduct X so high that additional process are needed to remove it?"
- Edit business letters and even brief memos for the appropriate tone and language.

The first two points are the most critical. Whether you are writing a two-line memo or a 40-page research report, you must have a clear purpose for writing. You must know to whom you are conveying this message. Then, you can put your message in the appropriate language and context.

Good editing will help you do this. However, editing to sharpen the focus of your writing is hard for many of us because we so enjoy the details of our work. If you are a researcher, save the details for research reports intended your peers. Even then, consider putting experimental details and theoretical discussions at the end of your report in an appendix.

Take a Tip from Newspaper Reporters

Journalism students learn "the five W's and the H" and good newspaper reporters use them every day. They are:

Who. Who did the work? Indicate this by identifying the authors. Criteria for authorship may differ in company reports and outside publications. Suppose a coworker needs to ask you an urgent question about the test procedures used to generate the information in your report and you are out of town. Technicians who performed the laboratory testing can provide this information.

◈ Reporting Failures

The old cliché is "everyone loves a winner." When you have a significant success, everyone wants to read about it. Your supervisor or team leader will urge you to write a report. A well-written report that describes a significant success can earn you and other members of your team new career opportunities and other rewards. Reading about successful projects can teach others how to succeed in their own work.

People often learn more from failures than from successes. Therefore, it is important to document failed projects, as well as successful ones. The word failure is a harsh and emotional term. Avoid it and other emotionally laden words. Report unsuccessful efforts in a matter-of-fact and unemotional way. Using the passive voice, normally frowned upon, can be useful to help avoid finger-pointing and blaming individuals for unsuccessful projects.

Avoid over-generalizing the failure. Sweeping negative statements can have an adverse effect on subsequent patent litigation, for example. Consider the researchers studying the pain-killing properties of furanoeudesma-1,3-diene. Comparing its performance to morphine, they found that ten times the morphine dosage was required to obtain equal performance by furanoeudesma-1,3-diene. While discouraging, these researchers avoided overgeneralizing and did not state the results indicate that all furanoeudesmadienes are less effective than morphine.

Clearly describe what was learned from the project that could be applied to other projects. For example, consider the marketing manager who reported the result of test marketing a new extra-mild baby shampoo in a midsized test city. The demographics were excellent and favored widespread use of the shampoo. However, parents disliked the product because it often left a gritty residue in the baby's hair. Subsequent research indicated that the same surfactant that made the shampoo so mild was incompatible with the very hard water in this city. This incompatibility resulted in precipitate formation, the gritty residue noted by parents.

Knowing the reason for the failure can teach useful lessons. The company should test the compatibility of all shampoos and bath soaps with the available water used in test markets. The company could consider marketing certain products incompatible with hard water only in areas where hard water is seldom a problem. For example, the baby shampoo could be made into a regional brand marketed only in major United States east coast and Great Lakes cities where water hardness in quite low. The company should not use a valued brand name on a test market product that might fail. Dissatisfied customers could associate the value with other successful products that bear the same brand name causing lower sales of the company's other products.

When identified as coauthors, the questioner knows who to call for information when you are unavailable. However, if these technicians merely followed your instructions in performing experiments without making independent contributions, it may inappropriate to include them as coauthors in a journal article (see below).

Why. Why was the work done? For internal reports, many readers already know the answer to this question. However, some will need to be informed and a concise statement of project goals is almost always valuable. Depending on the readership, the "why" can include both scientific and commercial reasons. The commercial reasons are essential to internal company reports and are often valuable in scientific journal articles, as well.

What. What was done? What do the results mean? A description of the experimental procedures or test design will answer the first question. The answer to the second question includes both the scientific or technical implications of results and their business implications. What are the authors' conclusions is usually the most important question the reader has.

When. When the work was done, the amount of time it took, and the time interval between writing the report and completing the work are often important for readers to know.

Where. Where the work was done is important. Do you draw your conclusions from laboratory, pilot plant, or commercial plant data? The answer to this question often indicates the degree of control you have over important variables and the analytical capabilities you have available. Scale-up factors and other variables often make pilot plant and commercial plant data more interesting and valuable to potential customers. Hence, where the work was done is often particularly important to readers of trade journal articles. The location often helps specify the exact equipment used to generate the data.

How. How the work was done often determines its validity and value. Hence, how the work was done and how the authors reached their conclusions are important for readers to understand.

Reporting Multiple News Items

What should you do if you have multiple news items to report? Consider the newspaper analogy. A newspaper does not put all the news into a single massive article. You should consider alternatives to putting all your news into one massive report. Since people tend to be intimidated by large documents, the longer your report, the fewer people will read it thoroughly and the more they may postpone reading it.

If the multiple news items concern different teams or departments, write separate reports. For example, suppose you attended an engineering conference on pipelines. You might write separate reports on the business and technical aspects of the conference. In the first report, you could describe discussions you had with customers and prospective customers. Send this to concerned marketing and sales representatives. Also send a copy to your supervisor. Describe technical aspects of the conference in a second report. If you cover diverse technical fields that are the responsibility of different departments or work teams, you may wish to write more than one report. For example, suppose your pipeline conference included sessions on new developments in the effect of soil properties on stress failures of pipelines and advances in metal alloys for pipeline construction. If you have a substantial amount of information to convey, separate reports on these two subjects may be warranted.

Tables and Figures

Data tables and figures can greatly aid readers' comprehension of your information. They can convince readers your data are valid and persuade them to accept your interpretation of the data and the conclusions drawn from it. Use data tables and figures appropriately. For instance, a fellow researcher may value your data tables and inspect them carefully to identify trends, look for experimental uncertainties, etc. However, large data tables may intimidate and confuse a salesperson or business manager. Therefore, determine whether a figure will work better than a data table in conveying the information.

Each data table or figure should have an objective and lead the reader to a single interpretation or conclusion. When you try to make a single table or figure do multiple duties and illustrate more than one point, communication problems often result. Tables and figures may become unnecessarily complicated. If you try to discuss two points simultaneously, the result may be that readers have a limited comprehension of both.

It is usually best to prepare your data tables and figures either before or immediately after preparing your outline. Preparing them before writing your outline may help focus your thoughts and aid in organizing your report.

Entire books have been written on the design of figures and tables. (See the Additional Reading section at the end of this chapter.) Some professional societies provide manuscript authors with guidelines for preparing figures and tables.

Writing for Clarity

Your readers will have one question when they pick up the letter, memo, or report you have written, "Why should I read this document?" The document title or the first line of text in a letter should answer this question. For longer reports, the author should place a short summary (100–150 words) at the start of the text. This summary and the report title will let readers know if they should read the document. The content of document titles, opening sentences of letters, and summary sections of reports should be both carefully chosen and examples of your best writing.

Poor writing can negate good choice of content and excellent document organization. Poor sentence and paragraph structure and inappropriate choice of words can have a devastating effect on effective communication. Some of the books in the Additional Reading section at the end of this chapter are excellent sources of advice on how to write clearly and concisely. Reading at least one will help even very good writers improve their skills.

The spelling and grammar checkers that are part of many word processors are helpful—so are writing-analysis software programs. These programs identify poor writing that may be grammatically correct, such as a very long sentence, and suggest improvements. However, all of this software is only a crutch and will not substitute for effective writing skills. In fact, using a writing-analysis program on a document can greatly increase the time spent writing. Far better to write well and use writing-analysis software to spot the occasional error or opportunity for improvement.

Writing for Publication

Once you master how to write clear, concise reports for coworkers and customers, you may want to consider writing articles for publication in a professional journal or trade magazine. Publishing well-written timely articles can have many advantages for both you and your employer. Good articles serve as advertisements of your technical and writing skills. They can convince readers that you and your employer have mastered a new technology, improved an old one, or developed new methodology. This can result in new business opportunities for your employer and career opportunities for you (either with your current employer or when you enter the job market). Professional recognition from respected peers can be a major source of career satisfaction.

Just as when writing internal reports, you must first decide what information you want to communicate. This must be information that readers value and will help them to understand better a specific technology or phenomenon. From the employer's point of view, this must be information that will lead readers to conclude that the employer is a progressive, innovative firm that understands the needs of its customers. Readers should be persuaded that the employer provides valuable products and services.

◆ Academic Publishing

Universities fit this definition of employers. Impressive scientific articles from its faculty convince faculty from other institutions that the author's university possesses valuable capabilities such as advanced instrumentation and the capability to use this instrumentation in innovative ways.

Professors from other universities may wish to collaborate with the author on future research or pay for access to some of the author's facilities. Students may decide to go to graduate school or conduct post-doctoral study at the author's university. The increased credibility that publications give faculty members aids them in obtaining additional grant money. The university will collect overhead fees from these additional research grants. Publish or perish is based on sound economics!

The decision to publish and determine what information to submit for publication is the next step you need to make, as well as determining where to publish. For industrial researchers, publishing in a respected research journal may be personally satisfying. Research or scientific journals cover basic research in various fields of science and engineering and are designed for specialists in those fields. Publishing in a research journal may be of little value to the employer if few potential customers read the journal in question. Hence, trade magazines serving customers' industries are often the preferred venue of publication for industrial scientists and engineers.

Readers of trade magazines are interested in keeping up to date on their industry's technology and learning information that will help them work more efficiently and their employer to operate more profitably. Writers for trade journals must keep these needs in mind when preparing their manuscripts. These will often be peer reviewed.

The same information may be written in different contexts and submitted for publication in trade magazines serving different industries. You can "customize" your manuscripts by adding information such as case histories pertinent to publication in one trade journal, but not another. For example, consider guar gum, a water-soluble polymer derived from certain plant seeds. This polysaccharide is an effective water thickener and finds applications in the food industry as an additive to ice cream, for example, and other products to provide a thick, creamy texture. Guar gum is also used to thicken water-based fluids used in hydraulic fracturing to increase production from oil and gas wells. Suppose a chemist or engineer has developed an improved process to remove extraneous material such as the skin of the seed from the guar gum. Perhaps a botanist has developed a strain of plant which produces a thinner-skinned seed. In either case, an improved guar gum containing less by-products is the result. Articles describing improved properties of this guar gum could be published in both food industry and petroleum industry trade journals. Each could include information that compares the general properties of the improved guar gum with conventional guar gums. This could include information pertinent to both industries such as the effect of various levels of dissolved guar gum on water viscosity. The article for the food industry could contain additional information such as the results of consumer taste comparisons of ice cream samples made with the old and the new guar gum. The article for the petroleum trade journal could contain information comparing the degree of rock pore plugging by solutions of the old and new guar gum. The results of oil well treatments comparing the old and the new guar gum is also an appropriate subject. In each case, the bulk of the information in the article is the same, although the terminology used is appropriate to each industry.

Remember when we said that employers will prefer publication in a trade journal that their customers read to publication in a scientific journal? This does not mean you must give up on publishing your work in a scientific journal. If you believe your work merits publication in a prestigious scientific journal, persuade your supervisor to let you write articles for both scientific journal and trade magazine publications. (You can use the oral communications skills discussed in chapter 4!)

Then, write two articles making sure each discusses the topic in an appropriate fashion for the different readership of scientific journals and trade magazines. The reader of a scientific journal article often has technical knowledge of a particular field equal to that of the author. Many readers of the trade journal article will not have such technical knowledge. Orient the paper intended for the scientific journal toward the laboratory research and include information and insights that would be inappropriate for trade magazine publication. Frame the scientific publication around the investigation and solution of a scientific problem. In contrast, you should frame a trade journal publication around the commercial advantages your solution to a problem has for its readers. Your trade journal article can contain scale-up studies and case histories inappropriate for publication in a basic research journal. Thus, the two manuscripts will differ in major ways.

Study well-written research journal and trade magazine articles to understand these differences. The authors' findings and conclusions are sometimes stated in an abstract at the beginning of a research journal article. The introduction is a statement of the problem, a summary of earlier efforts to solve the problem, and a discussion of why a new solution is needed. The authors then describe their own work. In contrast, the statement of the problem is often insufficient to persuade a trade journal reader to continue reading your article. Trade journal articles often open with a brief description of the authors' conclusions. This can persuade readers to decide that the article will be valuable to them so they continue reading it.

While specialists in other fields may read your scientific paper, science journal readers share a common concern that your experiments be soundly designed. You must convince them that you reach reasonable conclusions using valid methods of data evaluation and interpretation. They usually come from a similar research laboratory environment. In contrast, readers of trade journals are a far more diverse group. Some work in the industry itself while others work for companies that supply products and services to the industry. Some may be researchers; others may be manufacturing plant personnel, marketers, sales representatives, or purchasing agents. Your goal is to communicate with one or more of these groups on their own terms. Often, this means discussing technical matters in less-technical terms by avoiding the use of terminology that might be perfectly acceptable in a research journal. It may mean the author should use industry-specific terms that would be inappropriate for use in a research journal.

Many industrial scientists and engineers read both research journals and trade magazines. For example, many synthetic organic chemists working in the pharmaceutical industry will read both organic chemistry research journals and pharmaceutical industry trade magazines. This potential overlap of readership means that one should not rewrite the same article for publication in a science journal and a trade publication. Each article should contain a substantial amount of value-added information not present in the other. For example, the research journal manuscript can appropriately contain a detailed description of novel experimental or test procedures that the authors describe only briefly in the trade journal manuscript. The trade journal manuscript can contain a discussion of the commercial implications of the work often inappropriate for publication in a research publication.

After deciding to write a manuscript for submission to a particular journal, obtain a copy of that journal's "Author's Guidelines." These vary for each journal and include specifications on manuscript format and the preferred format for tables and figures. Many journals now accept (and some require) submission of manuscripts in electronic format, either on a computer disk or via e-mail. In these cases, the journal may specify the word processing software you are asked to use. One of the reasons I recently changed my software was because some magazines refused to accept computer diskettes prepared using my old word processing program.

USING COMPUTER TECHNOLOGY

Word processors are tremendously useful in the process of writing, revising, and printing an error-free manuscript. They can enable you to take the same report and tailor it for different audiences. For example, you could prepare a research report replete with data tables and discussions of both the commercial and the scientific implications of the work. Such a document would be distributed to researchers and stored in corporate archives to maintain a complete record of your work. Word processors make it relatively easy to modify your report for sales and marketing personnel. You could delete many or all of the data tables substituting verbal descriptions of results where appropriate. You could remove the technical and theoretical discussions and focus on the commercial potential of your work. Also, you could reorganize the report to tailor it for your readers.

If you only use your computer for word processing, you are barely scratching the surface of its capabilities. Using office suite programs, it is easy to import graphics such as charts, graphs, and diagrams into the text of your report and size them appropriately. Corporate intranets, the Internet, and e-mail make it much easier and more productive to collaborate with colleagues on joint reports and publications. Recently, I had occasion to collaborate with colleagues from two different companies on papers for two different trade associations. The collaboration with the colleague who had e-mail capabilities was far more efficient and less time consuming than the interaction with the other colleague. Much telephoning and pages sent back and forth by fax was necessary to collaborate with the colleague who could not be reached by e-mail.

Mastering some aspects of computer technology can help scientists and engineers work with sales personnel and graphics specialists to produce impressive, often animated, documents for use in company Internet home pages. Sales and marketing personnel also can give these sales presentations using a notebook computer, projection panel, and overhead projector. In small companies, the possibility exists of leaving out the graphics specialist and still producing presentations with a visual appeal that compares well to those of billion-dollar corporations.

Obtaining Corporate Approval to Publish

Before you start to write a manuscript, obtain approval in principle to submit it for publication in a particular journal. You must convince the appropriate individuals, usually your manager or team leader, and the manager of the relevant business unit, that it will be to the company's benefit to publish an article on this subject. The effects of publication on your employer's patent position and business must be determined. Sometimes this is difficult, often it is not. For example, consider the development of a new catalyst that enables your company to debottleneck a plant and increase production with little capital expense. This catalyst may not be patentable, or the manufacture and use of the catalyst may be such that it would be difficult to enforce a patent even if your firm has one. In these circumstances, it is most unlikely that the company would want to draw attention to its new catalyst and your request to write an article would be denied. In another case, suppose a newly developed drug passed animal screening tests with flying colors. It may or may not be to your employer's advantage to publicize these results prior to requesting permission from the United States Food and Drug Administration to evaluate the drug's effectiveness on human volunteers. Often, a business manager will present you or a project team with the idea of writing an article. In this case, business factors indicate that publication would benefit the company. A well-written and properly focused manuscript should face few problems in receiving approval for submission.

When you request permission to write a manuscript for publication, timing is important. It is usually necessary to have initially written an internal report so that your manager and fellow team members are familiar with the information and its implications for your company's business operations. If action on this information is required, it should be determined and implementation begun.

After you receive approval in principle, write your manuscript using the same approach outlined above for company reports. Have concerned coworkers review your manuscript before submitting it formally. They may have valuable suggestions for improvements. Do not ignore coworkers considered less technical, such as sales representatives. After using their comments to improve your manuscript, submit it to obtain the approval to send it to the journal. If your company has a submission form, fill it out carefully and completely. If you can include favorable comments from coworkers and team members, do so. For instance, suppose a company sales representative on your team reviewed your article and believes it would help persuade a potential major customer to purchase the new product described in your manuscript. This remark could aid you to persuade your marketing organization to approve your manuscript for publication. Circulate your manuscript to all concerned parties for their review. These reviewers could include your supervisor, the appropriate research and business managers, your patent or legal department, and your government regulations and environmental affairs department.

If you mention any customer's names in your manuscript, obtain their approval to do so before submitting your manuscript for publication. If you performed some plant trials in a customer's facility and do not mention their name in your manuscript, you may still need to obtain their approval. Why? Your description of the facilities could enable an industry expert to identify the plant where the test was performed and reach certain conclusions regarding your business relationship with this customer. Whether correct or not, you may prefer that these conclusions never be drawn. If you are confident that the description of the test facilities is too generic for the customer to be identified, you may omit obtaining their approval. The exception to this is if their approval was specified as a requirement in the agreement leading to performance of these plant tests. Clearly, this is a judgment call and requires careful consideration.

AUTHORSHIP

Properly including coworkers as coauthors can aid greatly in developing and maintaining productive and mutually rewarding working relationships. However, proper ethics must be observed. Coauthors must have contributed conceptually to the manuscript or participated actively in its preparation. For example, technicians who merely followed your directions in performing experiments do not merit being named as coauthors. You may wish to note their contributions in the acknowledgments section of your manuscript. However, technicians who participated in designing test procedures or independently managed a complex experimental program or field trial probably merit inclusion as coauthors. You can clarify any uncertainties by asking them to write the experimental section of the manuscript. Sales representatives sometimes collect testing information from customer plant trials of your firm's product or process. If the sales representatives write this portion of the manuscript, there is no ambiguity about including them as a coauthor. If analyses by others played a key role in the project, they, too, should be considered as coauthors. If they developed new or modified test methods or provided the interpretations used to guide the work and if these analyses are part of the manuscript, the analysts should be coauthors. However, if they solely performed routine analyses and you interpreted the results, mention the analysts in the acknowledgments section and do not include them as coauthors.

Fortunately, practices of questionable ethics such as automatically including research managers as coauthors have decreased greatly in the past quarter century and now seldom occur in most countries. Another judgment call comes when you receive the manuscript back from the journal with requests or requirements for revision. As author, you must decide if you should make the requested revisions. Then, having made the revisions, you must also decide if they are substantial enough to require that they be submitted for approval before sending the revised manuscript to the journal. This second decision depends also on your corporate culture and may vary from company to company. Consult with your coauthors on these issues. If you are uncertain about submitting the revised manuscript for approval, consult with your supervisor.

Also, consult with your sales representatives and marketing manager to determine if you should order reprints of your article for distribution to customers and potential customers. Usually, this is done when your revised manuscript is submitted. Waiting until the article is published is often too late.

MANAGING THE WRITING PROCESS

Knowing *when* to write is as important as knowing *how* to write effectively. Timely reports can be used as decision-making tools and thus are critical in project management. Prompt reporting also creates an atmosphere of progress in meeting project goals and serves to promote interest among corporate decision-makers. Timely reports can also enhance cooperation and enthusiasm among project team members. This type of report writing can be an important factor when managers evaluate your job performance. Timely report writing improves productivity and reduces stress caused by a backlog of unwritten reports.

Progress report writing can be a useful project management tool. Assembling and organizing data, describing results, and discussing their implications enables you to spot gaps in the data, poor data, or flaws in your reasoning. Additional lab work can often remedy some these problems. This is much easier when the equipment and personnel are still available, as opposed to waiting until the very end of a project when they may no longer be available. Alternatively, you and your manager can end an unproductive project sooner if timely reports indicate it will not yield useful results.

Timely reports can also assist in preparing sales aids. These can include product information brochures and advertisements, safety bulletins, premanufacturing notices, environmental impact statements, plant operations manuals, and other documents needed to commercialize or promote new products and services. Volunteer to participate in writing these documents or assist the professionals who do so.

Writing long reports (or a book!) can be very intimidating. Sometimes it may seem as if you will never finish. This feeling can be very enervating and subsequently delay the completion of the report. Reduce this problem by using your outline to divide the writing process into smaller units. When you finish writing each section, you can cross it off the outline. The sense of accomplishment will provide new energy for you to continue the report. Also, it will give you a better sense of the time needed to complete the report. Sometimes this can lead to the conclusion that you should divide a very long report into two smaller ones. This will enable you to get at least some information into your readers' hands in a more timely fashion. Also, a smaller report is less intimidating to readers and they may read your report more promptly.

It is easy to be trapped into spending evenings, weekends and even vacation time writing. You may feel you are stuck in a never-ending effort to catch up. However, sacrificing your limited leisure time to write reports is not the answer. Harvard economist Juliet Schor, author of *The Overworked American* (Harper-Collins, New York, 1991), has noted that longer working hours can actually reduce productivity. Therefore, extended working hours are not the long-term answer to the report writing dilemma.

The answer is twofold. You need to manage the report-writing process and your time efficiently. Consider report writing an integral part of your work. When beginning a project, allow time for report writing when setting schedules. Studying the project design should identify decision points at which managers and coworkers need timely progress reports. Consider the project completion date to be when your final report issues, not when you complete the lab work.

Set writing deadlines even if your supervisor does not. Robert Moskowitz, author of *How to Organize Your Work and Your Life*, recommends that you "estimate how long you expect a project to take and then schedule backward from your deadline. That will tell you when you have to get started." Be realistic when setting schedules. The writing process always seems to take longer than you think it will. Allow some time for slippage to perform additional experiments to fill gaps in your results or to verify data. By keeping a time log for several report writing projects, you can learn how long different types of reports actually take to write. This will increase your accuracy in predicting the time necessary to write future reports.

You now know when you should begin writing your report. However, do not use this date as an excuse to delay if you can start writing sooner. Plan both the report and the writing process before beginning the report. When you have a long document to write, first identify your readers and their information needs. Determine your goal in writing the report. What important information do you want to communicate? Is it to convince readers that a manufacturing process will be economically feasible, that a fluidized bed catalytic process is superior to other alternatives, that the results from animal testing warrant further development of a candidate drug?

Remember those laboratory reports from your introductory chemistry course in college or high school? Chances are your instructor required that the first section of this report be "Purpose" and you identified why you performed an experiment and the desired outcome of the experiment. It is the same with reports—whether about research results, plant evaluation of a new manufacturing report, or the description of a proposed marketing campaign. Your readers must know why they should invest the time and effort in reading your report. When you have clearly identified your purpose, the information you want to convey, and your readers, you are ready to prepare the outline of your report.

Outlining Your Report

Preparing the report outline is the most critical step in managing the writing process. Outlining organizes the report before you write the first paragraph. It aids in identifying reference materials that will be needed later. You should assemble these before beginning to write the report. Outlining can also help identify gaps in your data; gaps you can fill early without interrupting your writing later. (Doing more work after you thought you were finished can be both frustrating and stressful.)

FORGET ABOUT THE RULES OF OUTLINING

Do you remember all those rules about outlining you learned in grammar or high school? Don't worry if you don't. It does not matter where you use Roman numerals or capital letters. What matters is if your outline clearly indicates to you the content and organization of your report and how the different parts of it fit together. You may be the only person to see your outline anyway.

Also, do not use your outline as a procrastination tool to delay the start of your report. You can waste much time experimenting with the order of different sections of the report and making sure the outline exactly follows the rules learned in school.

Outlining divides your large report into smaller sections. What may have seemed like an almost impossible task becomes several smaller, more manageable ones. This makes scheduling the writing process much easier. Rather than dropping all other duties to spend days of work on your report, you should integrate writing into your daily activities. Your outline will help you use small blocks of time for assembling data, collecting references, or writing short sections of the report.

Outlining helps you write different sections of the report out of sequence. This allows you to begin writing a long report before completing all the experimental work. This can greatly reduce the time lag from finishing the lab work to completing the report. In addition, you can work on a section of the report while the pertinent information is still fresh in your mind. Waiting too long means you'll write more slowly and may need to spend additional time looking up notes and data.

Dividing the report into sections makes writing it much less intimidating. Consider each section an independent writing assignment. The satisfaction you feel in finishing one section helps provide motivation to write the next part of the report.

Do you travel a lot on business? Outlining also makes it possible to use your travel time more efficiently if you want to work on a report during a business trip. Outlining makes it easier to pack just the reference materials you need. When traveling, take enough materials needed for at least two sections of your report. This allows you to continue writing even if you meet an unexpected roadblock on one section of the report. You can work on less-technical sections such as the introduction and short sections while in the airport or on the plane. A quiet hotel room can be an excellent place to write longer or more technical sections of your report. Using a notebook computer will improve your productivity and the ease of revising your report.

Outlining makes it easier to delegate tasks such as assembling references. It is also easier to organize these tasks. Your librarian would prefer receiving one request for references and a generous deadline than several rush requests as they occur to you.

Outlining also makes it much easier for two or more authors to write reports. The outline makes coauthors clearly aware of their responsibilities and due dates. A chemist or engineer can assign sections of the report to technicians. Suitable activities to "farm out" include assembling data tables, preparing graphs and charts, and writing the experimental procedures section. Involving technicians in report writing promotes their professional development and improves teamwork. Teamwork enables you to reduce the time you personally spend writing the report.

WHEN NOT TO OUTLINE

You will seldom need outlines for short letters and memos in which you are delivering a focused piece of information. Standard formats for these documents can also help you eliminate the need to outline. For example, in a memo, always label your leading section "Purpose" or "Summary." Completing this section will help you focus your thought without going through the exercise of writing an outline. You can begin all your business letters with a subject heading that accomplishes the same purpose. Include in your report only the information that is pertinent to the subject. For example, your report on successful customer trials of your company's new product should not include news of the customer's planned reorganization unless it will impact their use the new product. If this reorganization will impact how your firm does business with this customer, it should be the subject of a separate report.

Once you have made sure your outline includes only information pertinent to the readers needs and your communication goals, determine whether you need any additional information. Often when preparing an outline, you may identify additional control experiments or an important issue that previously eluded you. In this situation, return to the laboratory or other workplace to perform the additional work needed to make your conclusions more convincing to your readers. However, be sure this work is truly needed and just not an excuse to interrupt or delay the writing process.

Other Techniques for Efficient Report Writing

Organize results and data as you generate them. The personal computer and spreadsheets can make this process much easier. Thus, when you sit down to write, the first step is not the tiresome task of assembling large data tables. Efficiently manage your writing time (see chapter 2). Your report outline can help you do this. However, you will also have to take control of your workday and make a concerted effort to change your work style. Get to work earlier. You can get up earlier and use the extra time to beat the rush hour traffic. By getting to the office early, you can sit down and write with few interruptions. If you live close to work, you could put in an early stint of writing at home before having breakfast with your family and then leaving for work. In either situation, solitude and the absence of interruptions will improve your productivity.

Determine when you are most productive and reserve some of this time for a daily undisturbed writing period. An old-fashioned daily work schedule drawn up at the beginning of the day or a microcomputer version can make this process automatic. Outlining will aid tremendously in scheduling report writing into your working day.

Write around missing information. Do not stop writing to spend thirty minutes searching for a reference or a week generating additional data. Schedule and complete these activities later while leaving gaps in your text or writing other sections of the report. Grouping activities can be key to increasing productivity and squeezing more time out of your schedule for report writing. For instance, make several phone calls in a row rather than scattering them through your workday. Accumulate materials and make one trip to the photocopy machine rather than several. Combine all your searches for missing references or data to a single library visit or one session with an on-line database. Have an early morning meeting with all the members of your project team to discuss the workday rather than meeting with them individually throughout the day. The key to productive writing is minimizing interruptions. Reserving time for writing and preventing interruptions becomes more difficult as people progress in their careers becoming involved in more projects and corporate activities. In his discussion on interruptions, Ernest Hemingway noted, "The telephone and visitors are the work destroyers." So, let your coworkers know your reserved writing time is off-limits except for emergencies. Have your secretary screen telephone calls, get a telephone answering machine, or use voice mail. If possible, avoid scheduling meetings during your writing periods.

The best days to reserve large blocks of time for report writing may be Monday and Friday. On Monday many of your clients, customers, and coworkers are not in full gear yet. Therefore, scheduling appointments and meetings for Monday is often a poor idea and you can reserve time for writing. On Friday many of your clients, customers, and coworkers begin to plan their next workweek and, perhaps, start to think about their weekend activities. Friday is a good time to write the letters, memos, and other writing assignments that have accumulated during the week.

Enlist your coworkers' help. Make sure they know when your regular writing period is. Reciprocate by not interrupting their writing time. Put a brief, humorous sign on your closed office door to let casual visitors know you are busy. Many writers generate their own unplanned interruptions. They rush off to find a reference, make a photocopy, or search for coworkers to ask them for information. Making these part of your grouped activities is more productive. Having taken steps to reduce unplanned interruptions, you should plan some. Productivity declines after working at a single task for more than about an hour. Planned interruptions can be a refreshing break that recharges your mental batteries and improves overall productivity. You can combine these breaks with chores such as making photocopies or cashing a check at the employees' credit union. Lunching with coworkers can be a welcome interruption. Eating lunch at your desk can make you feel virtuous, but often does not contribute much to productivity.

All these activities should reduce the need to write reports at home on evenings or weekends. However, when you do write at home, begin before or immediately after dinner. If you start late, physical fatigue will decrease your analytical capabilities, reduce your productivity, and weaken your commitment. Make sure your family understands that your work area is off-limits and they shouldn't interrupt while you are working.

One part of report writing that people often neglect is making revisions required by supervisors. Extensive rewriting decreases both your productivity and that of your supervisor. Rewriting is often the most difficult writing to do since you are already trapped in certain thought patterns. It can also be the most stressful part of the report writing process.

You can reduce the productivity loss and aggravation of rewriting. Show your report outline to your supervisor before beginning your writing. Make sure she finds your report organization acceptable. Carefully consider your readers before writing your report. What do they need to learn from the report? This must be your primary message. This message must not get buried in masses of information. For instance, often it is best to either omit the technical details that bench chemists and research engineers love or relegate them to an appendix at the end of the report. Learning your supervisor's reporting likes and dislikes is also important in reducing rewriting.

When you think you have finished your report, do not submit it immediately. Let it sit a couple of days. This aging process allows you to develop some emotional detachment from your writing. You can then review your report with a critical eye and make needed changes. For a major report, it is helpful to get constructive criticism from a trusted mentor or coworker. Only after making needed changes should you submit the report to your supervisor. Your report is truly finished only when you receive all the necessary approvals for it to issue.

When you finally finish that big report, reward yourself and express appreciation to others. Do something to make its completion a special occasion. This will make it easier to tackle your next big writing project. Take a break from writing for a couple of days. Celebrate by taking coworkers who helped on the report out to lunch. They'll be even more cooperative the next time you need their help in getting a report done. If you did a lot of writing at home, show your family how much you appreciated their cooperation in not disturbing your writing periods. Take them out to dinner, cook a special meal, or schedule a special activity.

Reducing your writing backlog and rewarding yourself can significantly reduce work-related stress. You will enjoy your job more and also be more productive. As you use your time during working hours more effectively, you will probably have more time to spend with your family.

ADDITIONAL READING

Bates, J. D. *Writing with Precision.* Acropolis Books: Washington, DC, 1986.
Beer, D. F., ed. *Writing & Speaking in the Technology Professions: A Practical Guide.* IEEE: New York, 1992.
Cain, B. E. *The Basics of Technical Communicating.* American Chemical Society: Washington, DC, 1988.
Chesney, T. A. R. *Getting the Words Right: How to Revise, Edit, & Rewrite.* Writer's Digest Books: Cincinnati, OH, 1983.

Day, R. A. *How to Write and Publish a Scientific Paper.* ISI: Philadelphia, PA, 1979.

Ewing, D. W. *Writing for Results in Business, Government, the Sciences and the Professions.* Wiley: New York, 2nd Ed., 1979.

Handbook for Authors of Papers in the Journals of the American Chemical Society. American Chemical Society: Washington, DC, 1967.

Murray, M. J. and Hay-Roe, H. *Engineered Writing: A Manual for Scientific, Technical, and Business Writers.* PennWell Books: Tulsa, OK, 2nd Ed., 1991.

Shulman, J. J. *How to Get Published in Business/Professional Journals.* Jelmar: Plainview, NY, 1988.

Strunk, Jr., W. and White, E. B. *The Elements of Style.* Macmillan: New York, 3rd Ed., 1979.

ADDITIONAL RESOURCES

Many professional organizations, including the American Chemical Society and the American Institute of Chemical Engineers, offer short courses on business and technical writing. Many colleges and universities offer evening courses on business writing, technical writing, and journalism. For those interested in improving their business communications skills, these courses are better investments than English composition courses.

4

ORAL PRESENTATION SKILLS

Effective public speaking skills are invaluable for a chemist's career advancement. Oral presentations play a major role in communicating research results and accomplishments. They are also a forum for decision-making: to continue or terminate a project, file a patent, commercialize a product, along with other decisions that may influence the speaker's career.

Today's chemists, engineers, and technicians are making more technical presentations than ever before: at project team meetings, at customer workplaces, at trade association meetings, and at scientific conferences such as ACS meetings. Of course, chemistry teachers rely on oral communication skills in the classroom.

Oral presentations may be relatively informal such as a one-on-one discussion with your research supervisor, a consultant, a customer, or a supplier. Presentations to your project team may also be informal. However, presentations to managers, particularly high-level executives, are usually quite formal even if the audience is small. Presentations before customer groups or at technical conferences are also formal. One possible exception to this is poster presentations. Even at large technical conferences, these are often informal.

TWO KEYS TO SUCCESSFUL PRESENTATIONS

Two keys to making successful oral presentations include defining your purpose and defining the needs of your audience. These determine everything about the presentation: organization, length, technical depth, and type of audiovisual aids. The first step is to have a clear goal. All oral presentations have the same general goals: to inform the audience, persuade them to accept your views, and to adopt your recommended course of action. Specific objectives depend on the subject you discuss.

The second step is to determine the needs of your audience. Who are they—researchers, executives, salespersons, students? What are their expectations and interests? To be successful, you must determine and meet their needs. For instance, consider a chemist who presents results and proposes to commercialize a new product. Business managers will want to know the economic significance of the results—what they mean in terms of investment requirements and profit potential. Sales personnel want to know the performance advantages of the product that will help them sell it. Both business managers and executives will need to know about any regulatory issues that might impact the manufacture and sale of the product. Technical details will most concern bench chemists. Engineers will want to know about the product's manufacturing process and what they will need to do to implement it. Plant production workers will want to know how the information you present affects their job responsibilities and activities. Safety and health issues will concern everyone.

Chemists, engineers, and technicians are increasingly making presentations to nontechnical audiences. Therefore, they must remember that what fascinates them may not greatly interest their audience. Consider the synthetic organic chemist who works for a pharmaceutical company and makes a presentation to a group of business managers and process engineers. The theoretical details of side reaction mechanisms reducing the yield of a drug may be exciting to the chemist, but be of little interest to her audience. They will want to know what the yield loss is, the impact on product pricing and profits, and what can be done to reduce the yield loss. After identifying the yield loss problem, the speaker should immediately discuss these concerns and save the reaction mechanisms for a different audience.

ORGANIZING ORAL PRESENTATIONS

At the beginning of all oral presentations, the audience should be informed of what they will learn from your presentation. This will give them a reason to listen to what you have to say.

You can organize oral presentations four different ways. The *problem-solving organization* is very useful when discussing technical problems before an audience of researchers. First, define the problem. Then, review possible solutions and their relative merits. Focus on the preferred solution and the results of its implementation. Reducing the level of technical detail and focusing on what the results mean in terms of producing and marketing a product will tailor this approach for a less-technical audience. Less-technical audience members include business managers, sales staff, plant operators, and (sometimes) customers.

Another way to organize your presentation is to present events in the order that they occurred. This *sequence-of-events organization* usually results in a less-concise presentation. Researchers and students sometimes use this type of organization in academic environments because it demonstrates their thought processes and problem-solving skills. For these reasons, the sequence-of-events organization can be a good choice for employment interview seminars, as well as graduate school seminars. However, be careful not to bore your audience with excessive detail. Avoid taking too long to get to the preferred solution to your problem.

The *analytical organization,* reporting what, how, and why is useful for determining the nature of a particular problem and reporting the facts of a situation. This is another useful approach when you are speaking before an audience of researchers. Judiciously used, it can also be the most effective organization for an employment interview seminar since it can illustrate how you go about solving problems.

Skilled speakers seldom use the *opposites organization* for technical presentations. This method uses a cause-and-effect or good news–bad news organization of information. This organization can be effective when presenting a series of alternatives or when you need to illustrate how you determined the preferred solution to a problem. The cause-and-effect organization can be useful for discussing very specific problems such as often occur in technical service work. For example, the effect of various process changes on a plant's production or product composition can be presented using a cause-and-effect approach. This is often the most effective organization to define the preferred solution to a technical problem. Carefully designed visual aids (see below) can be particularly important to the clarity of a cause-and-effect oral presentation.

You can use Toastmasters International's three-step approach:

1. Tell your audience what you are going to tell them.
2. Tell them.
3. Tell them what you told them.

This approach is effective with all four methods of organizing oral presentations. Speakers often begin this approach by using a slide to outline their presentation for the audience.

◈ The Cause-and-Effect Organization

This organization is best restricted to presentations made to fellow specialists. They can benefit from the information and your thought processes when later solving their own similar problems. Less-technical audiences will be interested primarily in "the bottom line"—the preferred solution. At most, failed approaches to problem resolution should be presented as a list. Briefly discuss the list (perhaps to justify how much time you spent on a project or forestall questions). Present these other approaches only after discussing the preferred solution.

Persuading the Audience

Whatever the general organization, begin by informing your audience why the information you are going to present is important to them. It is critical to seize their attention from the beginning. Unless it is relevant to the subject of your presentation, avoid beginning with a joke or anecdote. Many audience members lose interest when a speaker shifts from an irrelevant joke to the subject of the presentation.

You must persuade the audience to accept your information and adopt your recommended course of action. Present evidence, but remember facts seldom speak for themselves. Explain and amplify to be sure the audience understands both the information and its relative importance. Even if your audience is highly educated, they may have little expertise in your subject or their knowledge may be out of date. Determine this ahead of time. This way you can avoid the twin pitfalls of confusing an audience with advanced concepts or boring them with a lengthy introduction.

Induction and deduction are both useful in persuading the audience to accept nonobvious conclusions. Illustrative analogies can be very useful, but be sure they are both pertinent and appropriate to the audience. Citing opinions of respected authorities can be a useful means of persuading an audience to accept your point of view. Clearly establishing cause and effect can persuade an audience to accept your analysis and recommended course of action.

Most importantly, the speaker must explain the objective of the work. Often, this must be in such terms that a business-oriented audience can understand. Your relative emphasis on economic and technical factors should depend on the interests of the audience. They will have questions that need an answer. Do so in the detail and depth appropriate to their interests and needs. Recommending future courses of action is an important part of many presentations. Without future action that leads to economic benefits, business and sales personnel will consider most industrial chemistry R&D projects to be wasted efforts. When offering more than one course of action, discuss the pros and cons of each. Communicate the benefits that your employer will gain by following your recommendations; and be sure you can back up these claims.

Preparing Your Oral Presentation

Keeping the above considerations in mind, you are now ready to design your oral presentation. Organize it around the important points you wish to convey. Keep the time limit for your speech in mind. Experienced speakers may simply outline their presentations at this initial stage. You can also organize the presentation by writing it. When you have the content and organization of your presentation written as an outline or manuscript, try to anticipate audience questions and reactions to the various points you intend to make. Modify your presentation to include answers to likely questions.

EFFECTIVE VISUAL AIDS

The best-organized presentation can be a disaster if your visual aids are poor. The information you present and the size and interests of your audience should determine what type of visual aids you choose. Visual aids are of seven general types: 35-mm slides, overhead transparencies, slides presented on a laptop computer screen (which can be projected onto a large screen), televisions (TV) and videocassette recorders (VCR), flip charts, printed handouts, and the blackboard.

The familiar 35-mm slides are the best visual aids for large meeting rooms. Their most obvious disadvantage is the need for the room to be very dark so the projected slides are seen clearly. Overhead transparencies can be useful, particularly in small meeting rooms. You do not have to completely darken the presentation room for projected images to be clear. A lighter room promotes discussion during the presentation. By combining the use of overhead projectors, projection panels, personal computers, and graphics software, speakers can project dynamic, as well as static images. An alternative to the overhead projector and projection panel is a big-screen television.

A television equipped with a videocassette player can present dynamic images of an experiment or a process. Such presentations are best made before a small audience. Beside being a classroom tool, television is occasionally used at technical conferences. Some companies effectively employ a TV and VCR with a recorded presentation at trade show booths and hospitality suites.

For presentations to one or two people, some chemists use their laptop computers. This can be ideal for sales and technical representatives visiting a customer's office. By using notebook computers and with lightweight, portable overhead projectors and projection panels, speakers can present information to small groups at locations lacking audiovisual capabilities. When using a notebook computer, speakers can modify slides as their audience presents them with new information or "what if" situations. For example, consider a chemist explaining a multistep chemical manufacturing process. A business manager might ask the effect of a one cent per pound feedstock price increase on the profit margin of the final product. The speaker can enter the new data into an existing spreadsheet and immediately generate the requested information.

Speakers are increasingly using computers combined with projection technology to present visual aids in meetings of several or more people and in large lecture halls at conferences. However, many speakers have problems with such technology. Sometimes the technology does not work as reliably as desired. The sponsoring organization usually provides the computer and projection technology, while the speaker brings a diskette containing the presentation. Even if the correct software is on the computer hard disk, there may be subtle incompatibilities due to the version of the software used, a bad sector on the hard disk, or some other reason. Often, the speaker has not rehearsed the presentation using a computer or is unfamiliar with the operation of the presentation software. Computer projection problems are often lengthy and terribly distracting to the audience. Therefore, speakers should have a back-up projection method ready if computer problems occur.

In general, graphs or charts are preferred to tabular lists of data for slides or overhead transparencies. If your audience needs the numbers for analysis, distribute them as printed tables. You may wish to send this detailed information to them approximately one week in advance if some wish to digest the information before your presentation.

Consulting with more-experienced coworkers or with graphics experts can help chemists to design their visual aids properly. A particularly important and common problem to avoid is squeezing too much information onto a single slide or transparency. Another common pitfall is using dark colors against a dark background. These may project well in a small room, but be hard to see in a large conference hall. Graphics experts can provide advice to help you avoid this problem. Check your slides by projecting them in a room about the size of your meeting room to identify this problem beforehand.

Flip charts can be a useful alternative to overhead transparencies. They work well in small rooms and have the advantage that one can keep all the lights on during the meeting. They are usually hand-drawn and thus lack the polish of computer-generated graphics.

Poster Presentations

Attractive visual aids can be particularly important in poster presentations. Many organizations such as the American Chemical Society provide snacks and beverages at poster sessions. Many people stroll about the posters, often relaxed and casual with a drink or snack in their hand. They often search the crowd looking for a familiar face and merely glance at many posters.

At some meetings, you have very limited time to compete for people's attention. For example, the Technical Association of the Pulp and Paper Industry (TAPPI) holds their poster sessions during 30-minute intermissions from technical sessions. Attractive, attention-getting visual aids will help you compete in this distracting environment. Well-designed brightly colored slides accented by the use of a banner and very large lettering for your title can attract viewers for your poster. The same considerations important in preparing 35-mm slides and overhead transparencies apply to visual aids for poster presentations. In addition, avoid the use of glossy paper for your visual aids. Glossy paper can reflect overhead lighting and make an otherwise excellent visual aid hard to see from some angles.

Visual Aids for Cause-and-Effect Presentation

You can write on a flip chart or blackboard and gradually build up information such as a list of causes, effects, and costs. Consider a chemical engineer presenting his team's work on the effect of catalyst changes on the reaction rate of a process step in a drug-manufacturing process. He could tabulate his results as shown in Table 4.1.

Our chemical engineer may wish to emphasize that he found two methods of obtaining a three-fold increase in reaction rate in addition to focusing on product cost. To discuss the two issues separately while keeping the audience focused on the current issue, he can use the technique known as progressive disclosure. When using overhead transparencies, our chemical engineer could begin by covering the cost column with a sheet of paper. He then uses the first two columns of data to inform the audience that two different approaches increased the reaction rate three-fold. He can then pose and answer the question of which method is preferred on a cost basis by removing the paper from the last column of the table. Only now can the audience see the effect of catalyst changes on product cost.

When using an overhead projector and transparencies, this technique can be less distracting to the audience than repeatedly replacing a transparency with another containing an additional piece of information. Also, the simpler operation of moving an opaque sheet down a transparency to reveal additional information enables speakers to better focus their attention on the audience than the more complicated operation of replacing one transparency by another. Try using a properly sized cardboard sheet rather than a sheet of paper to help keep the desired part of the transparency covered.

You can also do this when revealing information using a series of 35-mm slides with each slide containing additional information added to a table or list. Our speaker can present the first two columns of data in one slide and the first and third columns of data in a second slide. Gradually revealing a set of facts or data can greatly improve the clarity of a cause-and-effect presentation organization. When presenting the above catalyst study to a group of research or plant engineers in an informal presentation, the speaker can forestall questions about costs until he has completed the discussion of rate effects.

TABLE 4.1.
Effect of Catalyst on Reaction Rate and Cost

Catalyst Change	Effect on Reaction Rate	Effect on Product Cost/pound
Replace Catalyst 1 with Catalyst 2	three-fold increase	$0.12 increase
Increase Catalyst 1 loading 25%	50% increase	$0.06 increase
Increase Catalyst 1 loading 50%	three-fold increase	$0.10 increase

The fifth type of visual aid is hand outs. These are best suited for small meetings or a presentation of numeric data that requires decision making. They are also best for presentations including a discussion between the speaker and the audience in which both must be able to refer to data presented in the talk. Consider the complete set of data for our engineer discussing methods of increasing the reaction rate of a process step in the synthesis of a drug intermediate.

The complete data set forms a complex table (See Table 4.2).

However, this is too much information to cram into one slide or overhead transparency even for a presentation in a small room. Separating the data into separate slides is difficult because some of the results are interrelated. If you are going to consider which method of increasing reaction rate is best, you need to consider both product cost and product purity.

The best approach in this situation is to keep the data in a single table and hand the table out to the meeting attendees. Be sure they have the information

TABLE 4.2.
Methods for Improving Process Yield

Process Change	Effect on Reaction Rate	Effect on Product Purity	Effect on Cost/pound
Use Catalyst 1, increase reaction temperature			
10°C	two-fold increase	no apparent effect	$0.02 increase
15°C	six-fold increase	amount of by-product 1 doubles, reaction stereospecificity to desired isomer reduced	$0.04 increase
Increase Catalyst 1, loading 25%	50% increase	no effect	$0.06 increase
Increase Catalyst 1, loading 50%	two-fold increase	no effect	$0.10 increase
Increase reaction temperature 10°C and increase Catalyst 1 loading 25%	three-fold increase	slight reduction in optical purity	$0.08 increase
Replace Catalyst 1 with Catalyst 2	three-fold increase	no effect	$0.12 increase
Increase reaction temperature 10°C and use Catalyst 2	10-fold increase	slight loss in optical purity	$0.13 increase

two days or so before the meeting to allow time to study it. Bring extra copies to the meeting just in case someone forgets to bring their copy.

Many speakers prefer blackboards or whiteboards for presentations before small groups in which there will be a lot of discussion and pooling of information. Flip charts may also be used for this. In this type of meeting, speakers often find themselves acting as secretaries writing down ideas or suggestions contributed by audience members. A disadvantage of blackboards, whiteboards, and flip charts is that speakers have to turn their back to the audience to write and thus find it difficult to maintain good eye contact with their audience. An alternative is to use a marking pen to write on a blank overhead transparency sheet while projecting it onto a screen.

REHEARSING YOUR TALK

When your visual aids are ready, practice your oral presentation. Prepare a list of obvious questions and appropriate answers. Prepare additional slides or transparencies to deal with these questions. Then, you will be prepared if these questions arise. For conference presentations and important business presentations, schedule a rehearsal in front of coworkers. Their input can help you identify and fix problems in your content, delivery, timing, and visual aids. If such a rehearsal is not possible, try rehearsing in front of a mirror. This is particularly helpful for speakers with audience eye contact problems.

Video- or audiotape a rehearsal session. Use your visual aids as if you were in front of an audience. This will aid in timing the presentation and in identifying trouble spots in your organization or delivery. Reviewing your tape will help you identify technique problems such as annoying speech mannerisms or poor eye contact with your audience. Recording your rehearsal is most valuable if an audience is present since you will tend to behave more naturally.

If you cannot videotape your presentation, ask for input from your rehearsal audience. Ask them to identify distracting habits that need to be eliminated

---

◆ Tips on Studying Your Videotape

Videotaping your rehearsal presentation will allow you to identify and eliminate problems:
- Avoid distracting movements such as twirling a pointer or tapping it against a lectern or table. Also, try to avoid annoying habits such as rattling coins in your pocket, touching your hair, or playing with jewelry. Dressing as you would for your formal presentation will help identify these habits. For

example, if you aren't wearing a tie, you will not fiddle with it in rehearsal, although you might do so in your formal presentation when wearing one.

- Use natural body language to emphasize important points. Do not over-dramatize. On the other hand, do not hide behind the lectern using it as a barrier between you and the audience.
- Be enthusiastic. If you do not appear interested in your topic, how can you expect your audience to be?
- View your videotape for evidence of poor eye contact with your audience. If this is a problem, keep eye contact in mind when making your formal presentation. If you are nervous, you may want to have a second rehearsal to concentrate on improving eye contact and eliminating other problems apparent in your rehearsal.

from your presentation technique. The night before your presentation, review your slides and the major points you want to make shortly before going to bed. Do so again early in the morning before your presentation. This double review can be quite helpful in fixing points in your mind and reassuring yourself of ample preparation.

FINAL PREPARATION

It is now the day of your presentation. Arrive before your audience and check the meeting room. Do the microphones work well? How loudly do you have to speak to be heard in the back of the room? Are the projectors operating properly? If you have time before the audience arrives, project your slides and transparencies to be sure they are in the proper order. If you are giving a talk at a professional society meeting, arrive early and introduce yourself to the session chairperson. She will appreciate knowing her speaker is present. A pleasant chat can relax you and help the session chairperson introduce you later.

Just before your presentation, calmly review the sequence of your main points. If you are nervous, recall how thoroughly you have prepared for your presentation. Remind yourself that the audience would not be there if they did not think you had something worthwhile to say. Another technique helpful to nervous speakers is to list each major point on index cards. Make these points coincide with the title or purpose of each of your slides or transparencies. Review the index cards the evening before your presentation and repeat this review early the next morning. Keep these cards with you during your presentation. Although you may not refer to them, they can provide a relaxing feeling of security.

DELIVERING YOUR PRESENTATION

After being introduced, walk briskly to the speaker's area. If you have a lectern, place your index cards on it. Take a deep breath and begin by decisively stating the purpose of your presentation. Tell the audience how you are going to accomplish this purpose before going into detail. The beginning of the talk is very important; it is then that you capture the audience's interest and establish control. This is critical to accomplishing your objective in making the presentation. A weak beginning will cause the audience to lose interest. Once this happens, it is hard to recapture their attention.

Once you capture their interest, maintain it. Be dynamic and forceful. Use body language, especially your arms and hands, to emphasize important points. Make good eye contact with members of the audience seated in different parts of the room. Do not fall into the common trap of staring at your own slides while you talk. Another way to lose good eye contact is to read a manuscript in front of your audience. Speakers who do so or recite a memorized text often speak in a monotone and lose audience interest as a result. Forgetting lines or

---

◈ Good Speaker Manners

Keep your presentation to the assigned time limit. If the session chairperson indicates you have used all the allotted time, finish as quickly as possible, ideally within two minutes. It is a discourtesy to both other speakers and the audience to go significantly over your time limit. This can throw the entire session behind schedule and make audience members miss papers they want to attend in other sessions.

Dress appropriately. In these days of relaxed business and conference attire, dressing up for a formal presentation is an expression of respect for the audience.

◈ Special Considerations for Technical Presentations to Nontechnical Audiences

Speakers must present information by using terms the audience can understand. Sometimes this means abandoning the technical terms that enable chemists, engineers, and technicians to understand each other clearly. This is not easy. Technical terms are useful shorthand for complex phenomena requiring lengthy explanation to nontechnical audiences. Keep the purpose of your presentation and your audience clearly in mind when deciding what technical terms to use. Use and define only two or three commonly used technical terms. A list of technical-term definitions at the start of your presentation can bore your audience; Defining numerous technical terms as you use them can disrupt the flow of a presentation. Your audience may focus on remembering definitions rather than following your reasoning to its conclusion.

Present the evidence that led you to your conclusions, but remember that the facts seldom speak for themselves. What may be obvious to a chemist or engineer may be a mystery to a nontechnical audience of sales representatives and business managers. Analogies can help your audience understand technical concepts. Make sure your analogies are both pertinent and appropriate to the audience. Citing opinions of respected authorities familiar to the audience can help persuade them to accept your views and conclusions; so can clearly establishing cause and effect. Pertinent anecdotes, especially humorous ones, can serve to make important but dry facts come alive and maintain the audience's interest.

Recommending a future course of action is often an important part of technical presentations for nontechnical audiences. Speakers should clearly show how a recommended course of action will solve a problem. Describe important implications of the solution, particularly economic ones. Should speakers offer more than one course of action, they should discuss the pros and cons of each. It is often difficult for a nontechnical audience to accept that there are few ideal solutions; real world solutions almost always have trade-offs. Clearly communicate the benefits your audience will gain by following your recommendations. Present these in well-defined terms: increased profits, reduced costs, higher production, decreased maintenance, etc. Quantify these benefits if possible.

Nontechnical audiences, having less intrinsic interest in technical subjects than chemists and engineers, will be less tolerant of poor presentation techniques. Therefore, time your talk carefully to avoid taking too long or having to rush through your conclusions and recommendations. Use charts and graphs to show trends, not technical details. Do not take the easy way out and use the same visual aids you would use for a presentation to a technical audience. Be sure your visual aids are free of technical terms and that they clearly communicate the points you want to emphasize. Graphs are preferred to tabular lists of data.

losing place in the manuscript can be devastating; a speaker may not be able to continue without a long pause that breaks the audience's concentration.

Use relevant anecdotes or examples to make your point. These can demonstrate the relevance of your results to your audience's concerns thereby maintaining their interest. Look at the audience while you speak and try to read cues to see if you are confusing or boring them. Actively involve your audience to maintain their attention. In a small audience, you can check for acceptance and understanding by asking the audience an occasional question. When you do this, try to pick out a member of the audience by name. Be sure you do not embarrass audience members or put them in a difficult situation. If appropriate, cite a member of the audience as an authority noting that the individual is present. If the audience is fairly small and you can hand out product samples or devices relevant to your presentation, do so at the appropriate time.

Suppose your presentation is informal and people feel free to ask questions. Do not let yourself become disorganized with questions that deal with points you will cover later. Do not let tangential questions make you lose your focus. A lengthy conversation with an individual about a minor point can cause other members of the audience to lose interest. Remember, it is hard to persuade the audience to accept your results, adopt your recommendations, or approve a course of action if they are not paying close attention.

Finish strongly; do not let your energetic flow of words wither to a weak finish. Remind the audience of your objective and summarize how you achieved it. Then, state your willingness to answer questions. When answering questions, do so concisely and directly. Do not respond emotionally to hostile questions or comments. If you do not know an answer, say so, but promise to find it. Do this promptly and convey this answer to the questioner and other interested parties.

DEVELOPING YOUR PUBLIC SPEAKING SKILLS

Effective oral presentations are a matter of careful organization, preparation, and self-confidence on the part of the speaker. These attributes plus practice can make almost anyone an effective public speaker. Watch yourself on videotape to improve your public speaking techniques.

Many corporations offer in-house public speaking courses for their employees. If your company offers such a course, take advantage of it. Toastmasters International has been helping people improve their public speaking abilities for many years; scientists and engineers could profit from joining this organization. Speakers improve with practice. Even if you are a good speaker, seek opportunities to make presentations so your skills do not become rusty. Treat each oral presentation as an opportunity to polish these skills and improve the perception that your managers, coworkers, customers, and peers have of your abilities.

ADDITIONAL READING

Oral Presentation Skills
Anholt, R. R. H. *Dazzle 'em with Style: The Art of Oral Scientific Presentation.* W.H. Freeman: New York, 1994.

C. R. *101 Secrets of Highly Effective Speakers: Controlling Fear, Commanding Attention.* Impact: San Luis Obispo, CA. 1998.

Kushner, M. *Successful Presentations for Dummies.* IDG Books Worldwide, For Dummies Press: Chicago, IL, 1996.

Rabb, M. Y. *The Presentation Design Book: Projecting a Good Image with Your Desktop Computer.* Ventana: Chapel Hill, NC, 1990.

Sinderman, C. J. *The Joy of Science.* Plenum: New York, 1985.

Interpersonal Communications

Haden Elgin, S. *The Last Word on The Gentle Art of Verbal Self Defense,* chapter 5. Prentice-Hall: Upper Saddle River, NJ ,1987.

Stacks, D. W., Padgett-Greely, M., and Hickson III, M. *Organizational Communication in the Personal Context: From Interview to Retirement.* Allyn & Bacon: Needham Heights, ME, 1997.

Other Resources

You can participate in well-established programs designed to improve your public speaking skills offered by the following national membership organizations:

Toastmasters International, P.O. Box 9052, Mission Viejo, CA 92690. Tel. (714) 858-8255. An affiliated group with the same address, Gavel Clubs, Inc., is for those who cannot participate in the full Toastmasters program.

International Training in Communication (formerly Toastmistress Club) 2519 Woodland Drive, Anaheim, CA 92801. Tel. (714) 995-3660.

5

Professional Networking Skills

Networking is another communication skill. It is more than just a technique to help job seekers find a position. Networking lets you access more scientific knowledge, technical skills, and career advice than available in your own department or company. It is an exchange of ideas, information, and resources. Carl Sinderman, author of *The Joy of Science,* defines networking as "the development of a mutually supportive, interactive, ever-widening circle of those with similar objectives, abilities, and views."

Networking is also about influence. Performing and reporting excellent science is the bedrock of becoming influential. However, networking can help good scientists become more influential. Influence translates into job opportunities, invitations to present papers, submit technical papers to journals, and

◈ A Networking Case History

At a recent ACS national meeting, I met someone peripherally involved in my new research area. She referred me to a fellow faculty member at her university with whom I initiated contact. This contact soon resulted in my employer supporting the university's basic research in an area of corporate R&D interest. As the liaison between my company and the university researchers, I have become acquainted with faculty members and graduate students. They have taught me new areas of expertise. The research review meetings have become an additional opportunity to network with my peers at other companies sponsoring this research. Additional networking at ACS and trade association meetings has resulted in joint research at a research institute and two universities. While no funding of outside research resulted from these contacts, the joint research has resulted in publication of three technical papers that I coauthored. These experiences have made me a strong believer in the value of networking.

write books. By organizing symposia, scientists can influence the direction of future research in a particular field.

Networking skills are not inborn; scientists and engineers can learn and cultivate them. Even introverts can master them to become skilled networkers. It takes years to build a good network. For long-term success, networking skills must involve giving, as well as receiving. Plan to spend some time and effort helping others in your network. When others help you, take the time to thank them. A simple thank-you note packs enormous power. Even if you thank someone orally, take time to also send a note. A thank-you note written on a humorous card will often find a place of honor on the recipient's desk or bulletin board. In return for exceptional assistance, send a copy of your note to your contact's supervisor. In-house e-mail makes this easy.

There should be an overlap of interests to make networking mutually worthwhile for the participants. However, in the career environment of the 1990s, it is unwise to limit yourself to networking only with specialists in your field. For example, industrial R&D bench chemists should include research managers and business managers in their networks. A good recruiter specializing in scientists and engineers or in certain industries also can be a valuable network member.

An exchange of research results and ideas (as appropriate in an industrial or academic environment) is an important part of networking. Scientific networking can result in opportunities to conduct joint research with colleagues from other institutions (see the sidebar above). Other rewarding opportunities that arise from networking include invitations to present papers at conferences, plan symposia, chair technical sessions, and contribute papers to technical journals and books.

As Sinderman notes, networking is a mutually supportive activity. Plan to spend some time and effort helping others with scientific problems, job hunting, etc. Giving is necessary to maintain and expand your network. As you establish yourself and widen your networking contacts, accept professional responsibilities such as reviewing technical manuscripts and organizing and chairing technical sessions at conferences. These activities pay off in building your professional reputation and adding to your networking contacts.

IN-COMPANY NETWORKING

Networking should begin in your own company. Your in-company network should extend across organizational lines. Sometimes your contacts will enable you to accomplish goals in spite of the system. In a large firm, contacts in other departments can be useful sources of information about in-house technical capabilities. For chemists at large research centers, contacts in an analytical department can be invaluable. These chemists can help you solve many problems in-house. Similarly, good contacts at contract analytical laboratories can be valuable.

Information scientists can also be valuable members of your network. Developing a personal rapport will help in discussions setting up the parameters of a literature search. Good networking contacts in graphics and photographic departments can also be very useful. Particularly if you work in a large organization, make the most of brief encounters. Try not to ask a vague question like "How's it going?" Ask coworkers pertinent, open-ended questions such as "How is work going on the new catalytic cracking catalyst?" By asking questions like these, even a brief conversation can be interesting and productive.

Attend company seminars and ask questions. At many companies, product managers are presenting more frequent seminars describing the status of their business. Attend these seminars and ask pertinent questions. As an analytical chemist or an information scientist, you may think you have little to gain by attending business presentations. However, you can learn more about your chemist coworkers' concerns and problems. For example, an analytical chemist doing surfactant analyses could benefit from attend a seminar on the state of the surfactant business.

NETWORKING AT PROFESSIONAL SOCIETY MEETINGS

Networking begins with your coworkers, but extends far beyond. Networking is much easier and more effective when you belong to organizations with members who have interests and goals similar to your own. Effective networking opportunities are an important benefit of professional society membership.

Professional society meetings offer excellent opportunities to make new contacts and add to your professional network. Your networking should be both local and national. Local meetings of ACS, AIChE, and other professional groups offer valuable networking opportunities. You can polish your networking skills for national meetings by attending local meetings. Also, you can make valuable contacts among chemists who work for other local companies and colleges. The meeting speaker is another possible recruit for your network.

Planning will increase your networking productivity. When you know you will be attending a conference, review the technical program and social events. Plan a daily schedule. With this in hand, call friends and acquaintances to see if they will be attending the conference. If someone you do not know is presenting a paper of special interest to you, contact this person as well. Make appointments for breakfast, lunch, or dinner. If meal appointments are not possible, try to meet at a social event or after an afternoon technical session.

Before a technical session begins, do not sit alone. Join a small group and try to enter the conversation. An alternative is to join a person sitting alone waiting for the session to begin. Strike up a conversation.

Name-dropping can be very useful in networking. If you meet Mr. Smith from XYZ Corp. and you have met Dr. Jones of XYZ, mention Dr. Jones' name to break the conversational ice. Shared acquaintances establish bonds. While you can exchange facts about third parties, do not gossip and spread rumors or highly personal information. For example, if Dr. Jones is not getting along with her supervisor, do not discuss this situation.

Even if your finances do not require it, sharing a hotel room with a colleague during a professional conference can be a valuable experience. You greatly deepen your acquaintance with your roommate. Lengthy, intimate conversations that are not otherwise possible in the hustle and bustle of a conference are possible in the hotel room. In addition, roommates can introduce each other to valuable contacts.

Conference poster sessions are excellent networking opportunities for both presenters and meeting attendees. They have become common at ACS national and regional meetings. Many trade associations are now experimenting with them. The ACS National Meeting SciMix is a particularly large poster session event. Attendees can discuss posters at length with authors and discover areas of mutual interest. Presenters can meet chemists interested in the same research area. Often speakers cannot do this in a conventional slide presentation. The audience may be large; the speaking room dimly lit even during the question period; or limited time available for questioning. These factors can inhibit some chemists from asking questions and limit networking. Members of the audience may rush off to attend a paper in another session. If the questioner and speaker manage to meet immediately after the paper, both may lose a valuable networking opportunity.

Ask intelligent questions after a presentation to help build your reputation and aid in networking with the leaders in your field. This can be particularly helpful to young technical professionals just beginning their careers and those in midcareer who have recently changed fields. However, asking questions for purely commercial reasons (trying to prove your company's product is better than the speaker's) or to try to embarrass the speaker is counter-productive. It demeans you as well, therefore only ask a question if you truly have one. Some professional societies require questioners to state their name and affiliation. Most ACS divisions do not. Questioners may be anonymous if the speaker does not already know them. This is not a problem in the one-on-one encounters common at poster sessions. If you note other questioners provide their name and affiliation, do so as well. This will aid other members of the audience if they wish to contact you later.

When attending a conference, participate in more than just the technical sessions. Taking a break from the technical sessions for hallway discussions or meetings over a cup of coffee can be very productive.

REMEMBERING NAMES AND FACES

Business cards are the indispensable tool of networking. Never be without them and be sure to hand them out at every opportunity. Ask for someone's card if

◆ Networking Is for Everyone

Networking is not just for the bench chemist or staff engineer. Government regulations specialists, safety professionals, patent attorneys, and others can discuss technical developments and news in their fields. Such discussions often take place under the auspices of various divisions of the American Chemical Society such as Chemistry & the Law, Chemical Health & Safety, Chemical Information, and Small Chemical Businesses.

Professional networking can also be very valuable for scientists and engineers who work in business functions such as marketing and sales. While they often cannot discuss sales to particular customers or various corporate strategies, they can discuss their opinions of the sales or business strategies of third parties. New technical developments by third parties and their possible influence on manufacturing or on markets can be rewarding subjects of discussion. Exchanging opinions about articles published in business and trade magazines can also be valuable.

they do not volunteer it. What if your contact has run out of business cards? Ask him to write his name, address, and phone number on the back of one of yours. When you promise information to someone or if they promise it to you, make a note on the back of their business card. That way you will remember the contact. Also, note any particularly interesting information about new acquaintances on the backs of their business cards.

Sometimes names and faces are difficult to remember. Using names frequently in a conversation is a good way to remember them. When he was at Hercules, Inc., Professor Ed Vandenburg (now retired from Arizona State University) would write down names and photograph people to help him remember both names and faces. Gordon Conference participants take home a conference photograph and name key to help them remember the people they met at the conference. If you remember a face, but forget a name, approach the person and greet them. If you are lucky, you will be able to read the name on his meeting badge. If not, remark that you remember meeting the person, but admit you have forgotten the name. This may be mildly embarrassing, but at least you have not missed an opportunity to renew a contact.

Another way to remember names and other information is to use a personal computer and a contact management computer program. Often used by sales personnel, these programs enable you to record extensive personal and professional information about an individual. Refer to this information before writing a letter, making a phone call, or having a meeting. This will improve the value of the interaction to both you and your contact. The disadvantage of this approach is that you must often spend time maintaining and updating your database. The initial transfer of your business card file of addresses and notes to your personal computer can be very time-consuming.

FOLLOW-UP

Follow-up is essential in maintaining your network. Fulfill commitments you make to people. As noted earlier, when people help you, be sure to thank them. Even in these days of e-mail, fax machines, and voice mail, a hand-written note is best. Telephone people in appropriate circumstances such as to discuss a recently published technical paper in a subject of mutual interest. E-mail has become a useful networking tool. Cards and short letters are useful contact maintenance tools. An end-of-the year letter can also be useful. This can be a business version of the holiday season newsletter many families exchange.

A healthy professional network requires continuing attention and maintenance. As your interests change, you may drop people from your network and add others. However, the opportunities for professional growth and knowledge it provides are well worth the effort.

NETWORKING RESULTS

What results should you expect from your investment of time and effort in networking? Sixty-four percent of the respondents to the 1993 ACS Employment Survey indicated their current job was a result of networking. Network to learn about job openings before employers advertise them. Many companies do not announce job openings, this way current employees can apply for them. In-company networking can alert you to these career advancement opportunities. Members of your network can also provide you with general information about employment opportunities in specific technical fields. Having versions of your resume and cover letters reviewed by members of your network can provide valuable suggestions for improvement. Some of your networking contacts can be a source of advice on how to handle specific job-related problems.

ADDITIONAL READING

Cohen, A. R. and Bradford, D. L. *Influence Without Authority.* Wiley: New York, 1990.
Peters, T. *The Pursuit of Wow!: Every Person's Guide to Topsy-Turvy Times.* Vintage: New York, 1994.
Sinderman, C. J. *The Joy of Science,* chapter 4. Plenum: New York, 1985.

6

CREATIVITY AND DECISION MAKING

Communication skills are the vehicle through which chemists, engineers, and technicians translate their ideas and knowledge into value for their employers. While superior communication skills enhance an employee's value to employers, so do their abilities to generate ideas and solve problems. Another valuable skill, particularly for managers, is decision making. Generating useful ideas, selling these ideas, and decision making are the subjects of this chapter.

GENERATING IDEAS

The process of creativity is not well understood. Knowledge is not creativity, although it can help fuel idea generation. Some of the chemists I have known with the best book knowledge of science were not highly creative. Many psychological studies suggest you can stimulate and develop your creativity. According to Louis Pasteur, "Inspiration is the impact of a fact on a prepared mind." There are many self-help books and short courses designed to stimulate your creativity. My personal opinion is that creativity is like typing; you can read about it, but, to become really good, you have to practice.

◈ Archimedes and Creativity

Observation creativity is the basis of one of the most famous legends of science. More than 200 years before the birth of Christ, King Hieron of Syracuse (a city in Sicily) asked the great scientist Archimedes to determine if his new crown was pure gold or alloyed with silver. Obviously, Hieron did not trust his goldsmith.

Archimedes was stumped until one day when he stepped into his full bathtub. The overflowing water led Archimedes to a new concept: an object immersed in water displaced an amount of water equal to its volume. He ran though the streets naked yelling "Eureka!" (I've got it!). After sorting out any problems with the police for violation of public decency laws, Archimedes got back to work. He immersed the crown in water and measured the water displaced. He did the same for an equal weight of pure gold. The water volumes were equal. If silver, less dense than gold, were in the crown, the volume of water displaced by the crown would have been greater than that displaced by an equal weight of pure gold. The lucky goldsmith presumably survived to make more jewelry for King Hieron.

There are two main types of creativity. The first type deals primarily with material things. This material creativity includes:

- Concept creativity—analyzing a problem from a new perspective and developing a new understanding of it, developing a new concept, and seeing connections between two apparently unrelated phenomena
- Product creativity—creating a new or improved product. Relying on concept creativity to do this is helpful. This concept creativity may be someone else's insights, not your own. (This is one reason why brainstorming is often so valuable.) Thomas Edison is a prime example of someone who relied on others' concept creativity to fuel his own prodigious product creativity.
- Observation creativity—seeing a phenomenon, particularly an unexpected one, and deriving a new understanding of nature or an idea for a new product or service from the observation. Observation creativity is a powerful reason for researchers not to rely solely on technicians to perform experiments. Some of my technician coworkers have received patents and become authors of technical papers because their observation creativity resulted in inventions and commercial products.

The second type of creativity deals with people and their interactions. This interpersonal creativity is as important in science and engineering as material creativity. Interpersonal creativity includes:

- Organizational creativity—figuring out new ways of organizing people to accomplish shared objectives. This includes assembling, managing, and working on project teams, organizing a symposium, and being an editor who works with authors to create a new book.
- Communications creativity—being able to express concepts and feelings in a way others can understand and accept. This includes writing and oral communications.
- Relationship creativity—being able to understand and relate with other people productively and in ways that satisfy your and their psychological needs.
- Inner creativity—self-understanding so you can deal with stress effectively and master inner fears and self-doubt. This inner creativity aids relationship creativity.

Problem Solving

In industry, creativity depends on successful problem solving. This begins with understanding and defining the problem. Often, chemists and engineers work alone in defining and understanding problems associated with their job assignments. They rely on their own reading and draw analogies from previous experiences. Sometimes they consult with coworkers or members of their professional network. Mentors are often particularly helpful. Senior coworkers and searching literature can tell you if anyone has tried to solve the problem before and what the solutions were.

Technicians will often consult with other technicians, chemists, and engineers. Increasingly, in today's corporations, managers are less available for consultations on specific problems. However, project teams can provide a valuable resource to help individuals define and understand their problems.

Often forgotten is determining when you have solved the problem. Some research programs continue longer than necessary because this question has not been answered. The answer to this question will help you conserve your resources and focus on high-productivity activities. To help answer this question, conceptually develop an ideal solution to your problem. Comparing previous solutions to this ideal can help you focus your problem-solving work where it will be most productive. By later comparing your solution to the ideal and to previous solutions, you can determine whether you need to do further work.

Brainstorming sessions can be a valuable tool in defining and understanding problems. Dr. Alex Osborn invented the process in 1941 and described it as "a method by which we can use our brains to storm creative problems in a way that keeps judgment from jamming imagination." Brainstorming increases the breadth with which you consider the problem. In addition to your own knowledge and experiences, those of others are applied to the problem. Common difficulties with brainstorming sessions include getting the appropriate group of professionals together around a table. Another problem is the tendency to start analyzing already presented ideas rather than focusing first only on generating

---

◆ Steps to Productive Brainstorming

1. Use a room with chairs grouped around a single table.
2. Appoint someone as scribe to write down ideas as the group generates them. Have flip charts or blackboards available. You want to be able to have everyone see all ideas as the scribe writes them.
3. Encourage an informal atmosphere. Have coffee or soda available.

4. Make sure the ground rules are clear. Inform the group that the initial period will be *only* for idea generation, not analysis. This initial period can be a set amount of time (at least 20 minutes) or until the flow of ideas stops. Note that only then will the group evaluate ideas.
5. Generate ideas as quickly as possible.
6. Generate as many ideas as possible.
7. Build on the ideas of others to generate more ideas.
8. Welcome all ideas.
9. After completion of the idea-generation phase, it is time to study and evaluate the ideas. This need not be done in the brainstorming session itself. However, it is often useful to consult with coworkers on particular ideas to get informed opinions. Consulting the literature can also help evaluate ideas.
10. As you collect information in step 9, note the advantages and disadvantages of each idea. Also, determine and note the type and amount of work needed to test each idea.
11. After completing step 9, prioritize the ideas. Base priority on the idea's likelihood of success, the amount of work required to determine the validity of each idea, and whether the needed resources are available to evaluate the idea. Input from the brainstorming group often is invaluable in this prioritization.

ideas. However, brainstorming sessions can aid greatly in the first step in problem solving: understanding and defining the problem. Brainstorming increases the scope with which you consider the problem.

There are many anecdotes about people who have awakened in the middle of the night suddenly with a clear solution to their problem. This is the phenomena of incubation. One famous example is Sir Isaac Newton sitting under an apple tree. It is doubtful that an apple falling on his head led to the theory of gravitation. More likely, the thought came to him as he was relaxing. Detaching your mind from a problem allows your subconscious to ruminate on it. New associations can occur resulting in a solution. According to Eugene Raudsepp, "Inability to relax, to let go of a problem, often prevents its solution." Hence, constantly thinking about a problem can often be counter-productive.

Illumination is the moment when inspiration strikes and an idea suddenly occurs to you. Preparation plus incubation can result in illumination. Illumination, the clear expression of an idea, is not the end of the process. Verification is needed to test and prove the idea is valid. Verification and the resulting development of a useful product, process, or service require much work. Often you have to persuade others to provide you with the resources to do this work.

SELLING YOUR IDEAS

The world will not beat a path to your door because you had an exciting research idea. In industry, an idea has to fill a need. Academic researchers, when preparing research grant proposals, need to provide evidence that success of your idea will fulfill a commercial or societal need.

Identifying your employer's needs is an excellent way to begin generating valuable ideas for industrial chemists and engineers. However, you will have to convince your supervisor, coworkers, and business sponsors that your idea is feasible and will generate profits for your employer. Failure to do this is the primary reason excellent ideas fall by the wayside.

◆ The Usefulness of an Idea Depends on Your Employer

Obviously, an idea for a new drug will be of little use to a petrochemical company. However, even within a given industry, the usefulness of an idea depends on your employer. Consider a pharmaceutical company. An idea for a hypertension drug that fills a gap in its product line could be very useful. Another company that already markets one or two excellent hypertension drugs has less incentive to fund the same product idea (unless its patents are expiring).

Your ideas also have to match your employer's resources. Suppose your employer would like to manufacture a new hypertension drug. A research proposal that, if successful, would require constructing an expensive plant may be of less use to a cash-short company than one having ample funds available.

This mismatch of the idea and available resources is another problem you must solve. An attractive solution is an idea for a manufacturing process requiring less capital investment. Your company might prefer a drug synthesis that needs a less-expensive plant, even if it requires more costly raw materials.

Do not forget, business managers can solve problems too. A creative business manager could work out a joint venture arrangement with a cash-rich company that would partially fund the expensive plant. Another alternative is to have customers for the product partially finance the plant. In return, they receive future business considerations such as guaranteed delivery of a specified volume of product at a specified price. (The cost of raw materials and energy could be factored into this price so the chemical manufacturer does not find itself in a ruinous situation when process costs increase.) This sort of business arrangement is becoming increasingly common. Another solution might be a joint venture with your employer contributing the technology and another company contributing most of the cash.

You have to tap the same creativity used to conceive your idea to sell it. The first step in selling your idea is to build it up. You would not enter a marathon without training to develop your muscles and endurance; you should not launch your idea into the cold, cruel world without adding some muscle to it. Begin by letting your idea simmer for a while on a mental back burner. This allows your subconscious mind to go to work developing creative insights and ramifications. Do not be too critical at this stage. The brain's critical functions often inhibit creativity.

After a few days, describe your idea in a dated "Note to File" or in your laboratory notebook. This will help ensure that you get credit for your idea. Then expose your preliminary, and possibly incomplete, idea to coworkers, mentors, and your supervisor for their thoughts. Veteran employees, in particular, may have valuable insights based on previous company efforts in related areas. The opinions of successful innovators are particularly valuable. View these critiques as opportunities to improve your idea.

Be sure you are not reinventing the wheel. If management rejected a similar idea earlier, find out why. The reasons for this earlier rejection may no longer be valid. If it still looks as if your idea has possibilities, research the public literature. See if your company's competition has put your idea into practice. Modern techniques of literature searching through computer databases make this relatively easy. Review trade journals, corporate reports, marketing information, patents, and technical papers for information relevant to your idea.

Armed with this information and your coworkers' comments, refine and clarify your idea. Try to turn it into a network of related ideas. Later, if one aspect of the idea does not develop as desired, other aspects may still be feasible. For example, suppose you are a process chemist with an idea for a new synthesis of a particular chemical. Can you synthesize other useful chemicals using the same process? Are the intermediates in a multistep synthesis commercially viable products? Can you convert process intermediates to additional commercially viable products? Answers to these questions can turn your ideas into a network of ideas. These ideas could result in a single plant manufacturing several related products thus making your employer less dependent on a single market or a single customer.

If you are in R&D, prove the feasibility of your idea by performing a few scouting experiments. If you cannot demonstrate your idea alone, enlist the aid of coworkers. Successful "bootleg" experiments are often the most critical sales tool in selling your idea.

Now is the time to evaluate your idea objectively. Be critical, but careful. It is very easy to kill a good idea prematurely. Get the educated opinions of as many people as you can: researchers, business development, and marketing personnel. If you can protect your employer's interests, get customers input. When offering your idea for others' comments, make it clear why you think the idea is worth pursuing. Have some reasonable suggestions about the resources and time needed to develop your idea. Be sure your idea fits your employer's business interests. For instance, a commodity petrochemical company is probably unlikely to implement an idea for a new fine chemical. Ideally, your employer is already successful in the markets where your product would be sold.

Your idea should also fit your company's culture. For example, suppose your employer is primarily a formulator and tends to avoid investing in relatively expensive chemical process plants. Whatever its technical merits, an idea for a high-pressure catalytic synthesis of a new chemical is unlikely to win much favor.

Your idea evaluation should result in a clear definition of what it is you will try to sell. You should understand and be able to explain why your idea is worth pursuing. Just as lack of muscle definition seldom wins bodybuilding contests, an idea without technical and commercial definition will seldom win management approval.

Be sure you feel strongly about your idea before you try to sell it. If you are doubtful of an idea's feasibility, that's usually a signal that the idea has a major flaw. You are now ready to begin selling your idea in earnest. First you must have a clear idea to whom you are selling your idea. According to Henry Ford, "If there is any secret of success, it lies in the ability to get the other person's point of view and see things from his angle, as well as your own." If you are a

◆ Preselling and Allies

Preselling and getting input from coworkers can give others a stake in the success of your idea and, thus, produce allies for you to sell your idea. One technique is to develop additional end uses for a single product. This often requires close cooperation with business managers and sales personnel. For example, suppose your company wants to market an eight-carbon alcohol. The intended end use is to synthesize plasticizers. Your scouting experiments indicate these plasticizers are cost-effective alternatives to commercial products. Suppose this alcohol could be a useful intermediate in manufacturing surfactants. A different division of your company manufactures surfactants. Discuss your idea with researchers and business managers in this division. They may want to evaluate surfactants made from your alcohol. Your alcohol now has a potential new market and you have corporate allies who support funding your idea. Your employer has an additional incentive to fund your work and commercialize the results.

novice, enlist the help of a more-seasoned coworker as your sponsor. This could be a mentor or your supervisor. Start by selling the "thought leaders" in your company. These are the people with a track record of successful innovation. Their opinions carry weight with corporate decision makers. Thus, their support can add greatly to your idea's credibility. Any salesperson will tell you "cold calling" is a difficult way to sell. Informally talk to the people to whom you are selling your idea. This "preselling" before making a formal proposal can be critical in tailoring the proposal to your employer's needs. Preselling makes it easier to anticipate questions and concerns when you formally try to sell your idea.

Your formal proposal should include both an oral presentation to the appropriate people and a written document. It is best to make your oral presentation first. Should someone raise important, but previously unconsidered, issues during your talk, you can modify the written proposal to deal with these issues. You will probably be selling your idea to technical and business managers simultaneously. Be sure to present your idea in terms your business sponsor can understand. The ideal presentation is one that takes the listeners to a point where they draw conclusions identical to your own.

Start your presentation with a "hook." Usually this is a clear statement of the business incentives to developing and commercializing your idea. Using appropriate "buzzwords" that fit your corporate culture is important. Examples include "proprietary technology," "improved feedstock utilization," "lower capital cost," etc. Ideally, your employer is already successful in the markets where your product would be sold. A product prototype can be a great hook. Arthur Fry used this approach to sell the 3M company his "Post-it" notes idea. He knew his product prototypes worked well as bookmarks in his hymnbook. They did not fall out and did not damage the pages when he removed them. Fry decided he needed some satisfied customers to persuade 3M managers to manufacture his product. He had more prototype "Post-its" made and gave them to the secretaries of 3M executives. They loved them. When they wanted more, he told them to call the marketing department. When they could not get them, the secretaries complained to their bosses. These executives were already sold before Fry made his presentation. You probably will not be as lucky as Fry. Your audience will be skeptical and have questions. Be direct in responding to these concerns. Any information you provide must be correct or you will damage your credibility. Some managers will want to be sure you have considered alternatives to your ideas. Be sure you can justify why your solution to a problem is better than the alternatives.

Be direct and honest, particularly in response to questions. Any information provided must be correct or you will damage your credibility. Ask for commitment. Make clear to your business sponsor what will be your next step. You may find you need to sell your idea in stages. Be politely persistent until you get a clear "yes" or "no" answer. Be prepared for disappointment. Sometimes the right answer is "no." However, do not let your idea fail due to inaction. If you receive approval to follow up on your idea, be sure to agree upon an action plan with your business sponsor. However, having an agreement does not mean that progress will happen automatically. Champion your idea and follow it through. Win the cooperation of those whose help you may need. Ideally, you will be asked to work on your approved idea. If your manager assigns someone else to develop your idea, find out why. Be sure there is a valid reason. Learning this reason may help you improve your job performance. Maintain an interest in the project if your manager assigns it to someone else. However, let it go; allow the person who has the project freedom to implement your idea and modify it as needed.

If management rejects your idea, keep it in your file. They are declining to spend the money to evaluate and develop your idea, not rejecting you personally. Your idea may be a good one submitted at a poor time. Changes in business conditions or advancements in technology may make it worthwhile to re-submit your idea later. Be sure to be diplomatic when you ask why your idea was rejected. Explain that you need to understand the reason why so it can help you when you submit future ideas. Avoid arguing again for your idea, it is the wrong time. Getting a clear reason for the idea's rejection will help you determine if and when it is appropriate to resubmit your idea. It will also help you modify the idea to improve its chances of acceptance later.

Selling ideas requires enthusiastic sponsors who are willing to take risks, plus a receptive and encouraging management. It also requires teamwork throughout the process. Teamwork gives you the best possible input during idea generation and definition. This increases both the chances of getting approval of your idea and successful development and commercialization. Be sure to give credit where due. There will be more than enough rewards to go around on a successful pro-

◆ The Politics of Selling

You will be selling your idea to different constituencies: R&D management, sales and marketing personnel, and manufacturing engineers. If coworkers will be working on parts of the idea, you will need to sell them too. Mentors can help young scientists and engineers identify the constituencies that have to be convinced that an idea is worth developing.

These constituencies have three questions. First, what are the chances of success? Second, what is in it for me? Third, what is it going to cost me? The answers to the questions will determine whether members of the different constituencies will support your idea. To sell your idea, you will need to convince the appropriate constituencies that the success of your idea is a reasonable prospect. In addition, you must persuade them that success will provide adequate rewards: a bonus or raise, a good performance review, improved chances for a coveted promotion or transfer, or the freedom to work on one's own ideas. Careful consideration is needed to determine the best idea-selling strategy for each person or constituency whose support you need. Mentors can help younger or less-experienced professionals develop these strategies.

ject. When trying to sell an idea, have realistic expectations. It is not considered a failure if you do not succeed in selling an idea, but it would be a failure not to try. Do not worry about submitting too many ideas. A large number of submissions does not necessarily mean more ideas will become R&D projects. However, it does mean that the quality of the ideas developed into projects will be higher. This can only be good for both you and your employer.

EFFECTIVE DECISION MAKING

Without effective decision making by you and your coworkers, your ideas will languish. Informed and timely decision making is an important key to career success, no matter what your job assignment. It is as true in research as it is in business management, marketing, and sales. Decisive people get ahead. Management author and consultant Peter Drucker notes, "A decision is a choice between alternatives. It is rarely a choice between right and wrong." The more decisions you make, the more you accomplish. If you are going to be wrong 25% of the time, it is better to make 100 decisions than 10.

Defining options is an important stage in decision making. You can rely on personal experience and knowledge. If you are a manager, you can also tap your staff's knowledge. Whether or not you are a manager, organizing a brainstorming session can generate alternative solutions to a problem. Decisions are not random guesses as to the best course of action; they must be informed. Therefore, before making a decision, you must collect the information you need to proceed intelligently. However, this does not mean gathering every bit of information on a subject. This can be an excuse to avoid or delay decisions. Decisions must be timely. Knowing when you have enough information to make a decision is very important.

Collecting information means mastering some of the techniques of on-line database searching, whether or not you perform the searches yourself. If you rely on an information scientist to search computer databases for you, you still have to understand the technology enough to choose appropriate key words. Even a relative novice can often suggest appropriate databases to search. Take a short course or independent study of the techniques of database searching to help you.

Effective business managers and sales people are well informed about conditions and trends in their own business and those of their customers and suppliers. Much of this information may be qualitative, not quantitative, in nature. For example, suppose your company manufactures dyes designed for polyolefin carpet fibers. Quantitative information important to your business decisions includes competitor's expansion plans, mergers and divestitures in the dye industry, and the development of new technology. Reading business and trade magazines and tracking patent activity can provide this quantitative information. However, what is important is not the information itself, but the impact of this information on your business. It is your interpretation of this impact that is important. For instance, quantitative information such as mortgage interest rates, the housing market, and wage trends play an important role in most people's decisions to buy new homes and redecorate current residences. Understanding these trends and their relative impact on the housing and redecorating markets can be critical in timing a plant expansion. Reading business magazines, studying survey results, and modest activities such as chatting with neighbors can provide our dye company executives with the information they need to decide on the scale and timing of their plant expansion. The constellation of many factors in decision making and the role of both conscious and unconscious thought processes can lead to that quality known as intuition.

The farther away your career takes you from the laboratory, the more your decisions will depend on psychological factors and "reading people." For example, during business negotiations, decision makers must know when the other parties have reached the limits of their ability to concede or compromise. How people react to information is more important than the information itself. Context often determines response. If we return to our dye manufacturer and consider the effect of mortgage rates on the demand for carpets (and thus carpet dyes), in some circumstances, an increase in the mortgage rate may lead consumers to delay buying a home in the hope the rate will soon decline. In other situations, such as occurred in the late 1970s, many consumers continued to buy homes despite high interest rates. Why? Because they expected mortgage interest rates to increase further (and for a long while they did). Even the laboratory scientist or staff engineer finds need of "people-reading" skills. They may not realize it, but they are often in negotiations. For example, in asking coworkers for favors, researchers must determine the most effective approach with each. Deciding upon the correct approach can be critical to your success in gaining the favor.

One must take an additional step beyond decision making: documenting decisions. Memories fade and your supervisor's knowledge may be incomplete. Therefore, when you make an important decision, document it with a memo. This will help assure that your manager and other concerned parties know who deserves the credit—and the rewards. Later, when you discuss your performance with your supervisor, refer to your memos to help you gain the promotion or salary increase you deserve.

RISK TAKING AND NEW PRODUCT DEVELOPMENT

Making decisions is a form of risk taking. More companies are emphasizing the importance of being willing to take risks. Much of this is lip service. However, educated risk taking is necessary for companies to grow. Risk taking is synonymous with exploiting new opportunities. Venturing into new technologies and new businesses, while it carries greater risks, also carries greater rewards. Even in older and, established businesses, new methods and directions can earn great rewards (while carrying increased risk).

Excellent chemists and engineers often work in businesses where technical opportunities—and risks—seem few. Yet appearances can be deceptive. Manufacture of polyolefins seemed like a profitable, but relatively mature business a few years ago. Then a wave of mergers and plant sales changed the players and the nature of the business. A higher premium is placed on being the low-cost manufacturer. Recently developed metallocene catalysts provide a means of producing improved products. The commercialization of new grades of polyolefins is intensifying competition.

Few of us are in careers where it is possible to avoid some risk taking. However, with enough information you can make these educated risks—risks you can calculate or estimate. Taking risks means being willing to take on tough assignments. However, take risks only after you have attempted to increase the odds of success and reduce the negative consequences of failure.

Consider a new product development effort. The original idea may come from any one of the team members. Having decided to develop this idea into a product or service, a business manager attempts to minimize the financial losses associated with the idea failing. Project teams can do this by designing the project to "front-end load" the risks. This means that most risks occur early in the development and commercialization process before the team has spent much money. Also, the team can increase the rewards by minimizing product-development time. Researchers can do this by working on appropriate activities in parallel, rather than sequentially.

To reduce risk and increase rewards, the business manager assembles a prod-

Development of New Drugs

One reason drug development is so expensive is that it is difficult to "front-end load" the R&D process due to the requirements of good medical practice and government regulations. Expensive human testing studies cannot begin until late in the drug-development process. As a result, commercially successful drugs must subsidize the costs of some very expensive failed programs. Only in an industry where customers will accept high prices could this process occur.

uct team. Members include sales and marketing people, business management, government and environmental specialists, process chemists and engineers, and product applications specialists. Plant production people can be part of the team. Later, appropriate specialists such as shipping and distribution experts and advertising specialists will join the team as the new product moves closer to commercialization. Thus, the team includes people from a variety of professional backgrounds. It is essential that all members of the team communicate effectively with each other. This often means avoiding the use of terms familiar only to specialists in one particular field.

The team's efforts begin with an assessment of market need. The sales people and applications specialists take the lead here.

Being Customer-Driven Is Not Enough

Being only customer-driven may be the surest way to drive yourself out of business. One must anticipate future customer needs and take the risks now to meet those future needs. Few potential customers saw the need for overnight delivery of packages and letters until companies such as Federal Express made such deliveries possible. Likewise, with overnight delivery available, few companies foresaw much need for the facsimile machine. Examples in the chemical industry are less dramatic, but the results they achieve are often equally impressive.

Consider the small-volume chemical user located far from the supplier's plant. Occasionally, production schedules and shipping times could make it difficult for the supplier to fill the customer's order in a timely fashion. The only alternative was either the customer or the supplier tying up assets in costly inventory. An alternative is for the customer to return the empty tote bins or drums when half of the latest shipment has been consumed. This lets the supplier know the approximate timetable for the next order and schedule production appropriately. In addition, the supplier can establish a system of dedicated totes or drums for particular products. Sometimes this can reduce or eliminate the time and expense required to wash totes and drums (and properly dispose of the wash water or solvents). The result is increased reliability in the manufacturing/shipment process for the customer.

This system was not the result of a customer request. The supplier developed it to solve what was seen as a potential problem: excessive inventory at both the supplier's and customer's locations. With the success of the system, the customer has awarded contracts for other chemicals to the supplier. In addition, the customer is demanding that its other suppliers use the same system. Our example supplier is also using this system to promote sales to other customers.

Suppose that managers have determined that sales and profits are sufficient to justify a product development program. Process engineers and chemists form a team to design the manufacturing process. Applications chemists work to determine product-performance requirements and the effect of impurities or by-products on performance. Government regulations specialists work with the process team, applications chemists, environmental chemists, and health and safety specialists to generate and assemble the information needed to safely and responsibly manufacture, ship, and use the new product. Should the Toxic Substances Control Act (TSCA) inventory not include this chemical, the appropriate testing must be done and an application submitted to add the new chemical to the inventory. Of course, technicians play a vital role in all laboratory work. At many firms, they will be team members. Their hands-on perspective can provide much useful input.

Project risks include inadequate product performance, a more expensive than anticipated manufacturing process, environmental or safety concerns that will limit the potential market, and the cost of the testing required to list the chemical on the TSCA inventory. One important consideration often neglected at this stage of the process is the physical form of the product that the customer prefers. For example, customers who are used to handling liquid products may resist using solids or slurries. Providing a high-performance economical product in an undesirable form can greatly limit its market acceptance.

Assuming no insoluble problems arise and updated economic projections remain encouraging, pilot plant or small-scale commercial production begins. Product supplies should be adequate to meet the customers' needs for samples to perform their own laboratory or small-scale plant tests. Applications chemists and technicians should work with sales staff presenting information on the new product, its properties, and advantages to customers. Customers will also want information on how to handle the product and on safety concerns and environmental properties. The manufacturer's team should also seek input on what additional information the customer needs. Applications specialists will perform the additional lab work needed to provide this information. This may include bench-scale tests of the performance of the new product in the customer's process. Success here can lead to customer plant trials using plant-manufactured material.

Before customer plant trials, government regulations specialists will need to prepare a material safety data sheet (MSDS) for the new product. Risks at this stage of the development and commercialization process include unexpected difficulties in manufacturing scale-up, lower than expected performance from the plant-manufactured product, and undesirable physical properties not predicted from laboratory experiments. For example, small amounts of by-products produced using a commercial grade of catalyst may result in a liquid product being light yellow when it needs to be clear and colorless. If they cannot prevent this problem, plant engineers may need to add a decolorization step to the product synthesis. This would increase product costs. Another risk is that the new product may fail in customer plant trials. It may not deliver value-added performance that will persuade the customer to purchase it. Either situation requires a major reassessment and lead to either major redesign or termination of the project.

Success in customer trials is often the trigger for the rapid preparation of promotional materials by members of the sales and advertising teams. These are usually trade journal advertisements and product brochures. Applications specialists may write technical papers and case histories describing the improvements delivered by new product. The supplier may rent exhibit booth space at major trade shows to further advertise the new product.

Other risks are present throughout the new product development and commercialization process. The product or process in which the new product is used may become obsolete, thus drastically reducing the potential market. Competitors may beat you to market with a similar product. Even if you continue and commercialize your own product, your market may be smaller than anticipated while competition reduces profitability.

Every member of the product development and commercialization team shares in all the risks described above. However, they do not share equally. For example, the risk taken by environmental and safety specialists is negligible unless the initial assessments they provide are faulty. Chemists, engineers, and technicians take considerable risks associated with technical issues. The marketing and sales specialists assume risks associated with their perception of market need and acceptance. The business manager accepts risks associated with his responsibility to spend company money and resources wisely. The more one could anticipate the risks and the more closely project failure is due to a specific risk factor, the greater the impact on the individual who undertook the risk. However, the same is true of the rewards. Many bench chemists, engineers, and technicians jealously look at the rewards garnered by project managers and business executives. These are the people, however, who take the greatest risks on development and commercialization projects. The negative consequences of failure are usually highest for them, as well. Consider bench chemists and staff engineers. If their supervisor concludes there was no way they could have anticipated a technical problem causing failure of a project, they will suffer little, if at all, as a result of a failed project. The same is true of technicians working on the project.

REDUCING THE DOWNSIDE OF RISKS

You can do much to reduce the negative consequences of project failure. Continuing our example of a new chemical product, applications specialists could develop additional uses for the product. A problem that limits product acceptance in its primary market may have little effect on its alternate uses.

Let us return to our problem of product color. We wanted a clear product, but the plant produced a yellow product due to the commercial grade of catalyst used. Process engineers decided to install a decolorizing step to produce the colorless product needed for its primary application in a formulated consumer product. This additional process step increases manufacturing costs. The manufacturer can absorb this higher manufacturing cost, thus reducing profitability. Alternatively, the manufacturer could pass along all or part of this additional cost to the customer. However, the higher product price could reduce sales. Suppose an applications chemist identifies another use in which the yellow color is not a problem. The manufacturer could divert some product to this market and only decolorize some of the product, thus improving overall plant profitability. Your employer could limit any decolorized product price increase and provide increased sales to two markets. Alternatively, the process chemist could find a way to manufacture another chemical from the process that is producing the yellow color in making product 1. Thus, even if the manufacturer or customer encountered problems with product 1, success with the second product could enable the project to go forward.

Proposing these options and putting appropriate plans in place is best done early in a project when judgment is not clouded by the stress associated with impending failure. Also, higher-level management may become so negative about an impending failure, they may not want to hear any more about it. Therefore, it is best to define and communicate these options early.

These options should include scenarios for both project success and project failure. For instance, consider the second small-volume application mentioned above. If product 1 fails in its primary end use, you may wish to license manufacture and sale of product 1 for the small volume application or manufacture of product 2 to a specialty chemical company. The licensing fees can reduce the cost of project failure. If product 1 is commercially successful in its primary application, you may still want to license it to a specialty chemical company to supply the alternate small-volume end-use market. However, you could manufacture this product and ship it to the specialty chemical company or a distributor. They would repackage it appropriately (transferring it from tank truck to totes or drums, for example), and then sell it to a market of numerous small customers the commodity chemical company is poorly organized to serve.

The more realistic options you can develop for a project, the more likely some will succeed. Identifying these options early in the project will help avoid premature or ill-informed decisions later when the project team encounters problems.

ADDITIONAL READING

General

Drucker, P. *The Effective Executive*. Harper & Row: New York, 1966.

Grossman, S. R., Rogers, B. E., and Moore, B. R. *Unlocking Creativity in the Workplace*. Wordware: Plano, TX, 1988.

Megill, R. E. *How to be a More Productive Employee*. PennWell: Tulsa, OK, 1980.

Peters, T. *Liberation Management*. Alfred A. Knopf: New York, 1992.

Strasser, S. and Sena, J. *Work is not a Four-Letter Word*. Business One Irwin: Homewood, IL, 1992.

Intellectual Property

Foster, F. H. and Shook, R. L. *Patents, Copyrights, & Trademarks*. Wiley: New York, 1989.

Gordon, T. T. and Cookfair, A. S. *Patent Fundamentals for Scientists and Engineers*. CRC: New York, 1995.

Mosley, Jr., T. E. *Marketing Your Invention*. Upstart: Dover, NH, 1996.

7

INTERPERSONAL SKILLS

No one works alone—not even the most brilliant ivory tower theoretician. We all rely on others to help us accomplish our goals. To do so we have to interact with them in congenial and productive ways. In this chapter, we will discuss how to work productively on teams, resolve workplace disagreements, and deal with workforce diversity. We will also discuss how to ethically influence others to help you. Since interpersonal skills are also key to handling promotion issues, this subject will also be discussed in this chapter.

TEAMWORK SKILLS

Today, an increasing amount of work is done by teams. This is true both in industry and in academia. Technical professionals have to coordinate their work with that of their coworkers and work with them in a cooperative, amicable manner. This has long been true for chemists and engineers who work outside R&D in chemical manufacturing plants or in sales, marketing, and other business functions. Much industrial R&D is being performed by teams that include researchers from different R&D departments and participants from business development, production, sales, and government regulations departments. (The trend to form teams is also advancing in academia. For several years, the mantra has been interdisciplinary research encouraging academic scientists from different departments to work together.) While some individuals question whether an R&D team can be as creative as a highly talented individual, it appears that workplace teams are here to stay, even in R&D.

Industrial researchers have long had to work together in a limited sense. A chemist would develop a new product or synthesis. He would work with analytical chemists to determine chemical structures of products, intermediates, and by-products. Should the company decide to commercialize the process, he will work with engineers to develop the manufacturing process while applications chemists study possible uses of the product to create markets for it. For many years, the contact between synthetic chemist, research engineer, plant engineer, and applications chemist was limited. This has changed. In today's workplace, scientists and engineers in many companies work in teams that manage a complex assignment such as developing a new synthesis or product.

Working on a team is only one of four relevant workplace trends. As companies work more closely with their suppliers and customers, chemists and engineers must work more closely with supplier and customer personnel to resolve or prevent problems and develop improved products and services. Downsizing is another trend that has forced coworkers to work more closely together. For example, downsizing means analytical departments have fewer resources. Analysts must cooperate with coworkers who submit samples to decide the most economical way to answer the technical questions that need to be answered. Working together, they must determine priorities and completion dates in coordination with the analysts' other work. They must resolve any disagreements over completion dates. All participants must do this in a professional, cooperative way (see below). Another relevant trend that increases the importance of teamwork is outsourcing. As different functions are outsourced, chemists and engineers must often consult with the people who perform these functions to be sure assignments are performed properly and on time.

WORKING ON TEAMS

More and more companies are embracing the team approach to business. As a result, many professionals—researchers, marketers, sales people, plant personnel, and managers—will have to learn how to work productively on interdisciplinary teams. This is not always easy. An entire industry has grown up around improving team performance. Team leaders use team-building exercises and outside facilitators to analyze and improve team performance. Each individual's interpersonal skills become more important by working on teams. Leadership skills will be by persuasion and personal example rather than by merely issuing orders. For example, employee empowerment has modified the "command and control" nature of most chemist—technician working relationships. Now chemists and technicians may work together with fellow team members who have different job responsibilities.

Employer efforts to reduce product-development cycle time are a major factor in promoting the use of teams. This has resulted in many researchers, plant chemists and engineers, marketing specialists, government regulations experts, sales personnel, and others who are working on the same project team to coordinate their efforts to achieve greater productivity. Even sales people who call on customers must be team players. Often they work with their manager, marketing managers, manufacturing and shipping personnel, and, in many industries, a research and product development staff who provide customer technical support. Employees who get ahead are those demonstrating that they are productive team players.

So how can you do this? It takes a systematic approach that should begin the first day you start on the job. If consistently followed, these ten steps, will establish you as a valuable team player.

1. Adapt to the corporate culture without sacrificing your individuality. Dress in a similar fashion to coworkers in the same classification or in the fashion of those on the next higher rung of the corporate ladder. If the mode of dress for your position is white shirt and tie, do not wear a T-shirt and jeans to work. Conversely, do not wear a suit in a "roll up your sleeves" casual work environment. If the custom is for coworkers to go to lunch together, join them even if it means staying thirty minutes late to get your work done. Do many of your coworkers go out for a beer after work on Friday? If so, join them even if you do not stay long and even if you do not drink. The important thing is relationship building, not the beer. However, do not spend so much time socializing during working hours that it detracts from either your job performance or your supervisor's opinion of you. Be friendly at business meetings, but once the meeting begins, stop socializing and concentrate on the subject at hand.

2. Analyze your supervisor's work style and likes and dislikes. Unlike the first step, this takes time. Observe your supervisor's behavior in meetings and how he does his job. Listen to your coworkers' opinions of the boss, but do not blindly accept them. Observe and form your own conclusions. As you learn, modify your work habits to his. For example, if your company has flextime working hours, determine if your boss is an early bird or tends to work late. Adjust your work hours to conform to his. If he does not like to read long detailed reports, keep your reports short and convey details to him in a face-to-face meeting.

3. Follow the chain of command within your company. Nothing will annoy a supervisor more than a staff member who goes over her head with complaints or a request to alter one of the supervisor's decisions. Chances are the first thing the higher-level manager will do is back up your supervisor. You will both annoy your supervisor, who naturally resents your action, and the higher-level manager. Your supervisor's manager may become concerned that there is a problem (you) in the supervisor's department. So what can you do if you do not like a course of action your supervisor is going to take? Do not wait until the decision is made. Discuss your point of view in a logical unemotional manner. Depending on the situation, it may be best to do this one-on-one with the supervisor. If most staff members feel as you do and the decision concerns them as well, bring up the matter at a department meeting. If you do this, be sure the supervisor knows that the matter in question will come up so she is not "blind-sided." Once a decision is made in a meeting with the supervisor, keep quiet and do your best to implement the decision. If the decision is something you violently or morally disagree with, do not sabotage your department. Consider asking for a transfer or getting a new job.

4. Be a loyal member of your supervisor's team. Be an active participant in meetings. Avoid criticizing your supervisor to outsiders. Support him in

discussions with his supervisor. Do your part and assist others in achieving departmental goals. Failures in a particular department reflect on all department members, not just the supervisor. Do not go to your supervisor with problems only. Try to develop possible solutions or options to discuss with him. That will make you look good and makes his job easier. However, do not try to ignore problems in hopes that they will go away. Discuss them with your supervisor and, if appropriate, with coworkers. People enjoy solving problems. Accept your department's goals and work toward them in a cheerful and enthusiastic manner. A happy group of people is more productive than a department going about their job in a resigned or reluctant fashion.

5. Cooperate with coworkers. Try to assist coworkers when they request advice, help, or information that concerns their job assignments. Also, do not be afraid to go to them when you need advice or information. However, make sure that assisting others does not interfere with fulfilling your own job responsibilities. There is a difference between being a team player and having a coworker take unfair advantage of your team spirit. Though you do not have to like every member of your department, being civil and working together productively will make your supervisor's job easier and yours more pleasant and less stressful.

6. Be Mr. or Ms. Reliable when fulfilling commitments and meeting deadlines. Of all personal qualities, including brilliance, reliability is the one that managers prize most. Finishing a project on time or early can give your work an extra glow that it otherwise may not deserve. Do not make commitments you cannot keep and do not dodge responsibilities you can easily handle. Prioritize your work assignments. Consult with your supervisor to be sure he agrees with your assessment. Then, do not allow high-priority projects to slide without your supervisor's approval. Always keep your manager informed of your progress on projects. In doing so, if difficulties occur that may interfere with a deadline, your supervisor will be less annoyed and more understanding.

7. Show initiative. Share your ideas at meetings or in discussions with your supervisor and coworkers. If you have a suggestion concerning another department's area of responsibility, make sure that you understand the problem and that your solution is a good one. Consult with your supervisor on how best to communicate the information to the other department since suggestions from outsiders are often resented. If possible, write a diplomatic memo that describes your idea. Be sure a copy goes to your supervisor. This can often open up a new avenue for career advancement. The other department head could become a new advocate for you in management meetings. Another way to show initiative is to become an expert in a new technology. This may be a new computer program, a new technique such as personal Web page design, or a productivity-enhancement tool adopted by your company such as activity-based management. Becoming an expert in a new technology can present new opportunities for career advancement. However, do not spend so much time solving other people's problems that you neglect your own. Do this too often

and, when you approach your supervisor with suggestions concerning another department, you will get concern about completing your assignments rather than a pat on the back.

8. Improve your competence in job-related skills. Learn to use helpful computer software such as word processing, spreadsheet, database, graphics, and statistics programs. Be alert to opportunities to apply this technology to your work. Be willing to help your coworkers and your supervisor in the areas where you develop a special competence. This can give you a value to the company beyond that of your formal job description. Such value can improve your chances for promotion or transfer while improving your job security. And also, as in step 7, avoid becoming so involved in an area of special competence that you neglect your ordinary responsibilities. Exceptions to this are when you decide to switch fields or pursue an alternative career. Another exception is when the time required to gain special competence is not excessive.

   Working in teams requires increased oral and written communications with team members. Periodic reports, either formal written documents or informal e-mail messages, will keep fellow team members informed of your progress. Often you will be making oral presentations at team meetings. Therefore, written and oral communication skills (see chapters 3 and 4) are becoming increasingly critical. For example, job hunters are evaluated on their written communication skills as demonstrated by their resume and cover letter and their oral communication skills during job interviews and telephone conversations.

9. Take advantage of in-house courses to improve job-related skills such as business writing, public speaking, and effective listening. Many companies devote a lot of effort to create useful in-house courses. Since they are often tailored to company needs, they can be superior to those offered outside the company (as well as much cheaper for you). Local sections of professional societies often have audio and video courses available for loan to members at a modest fee. Take advantage of these and study appropriate subjects. They offer the advantages of home study and flexible hours. Generally, audio courses are better suited to technical subjects in which class discussion is not of great importance to the learning process. Classroom situations work best for subjects in which discussion is an important part of the learning process. If you can commit the time, consider taking night or weekend courses in technical or job-related skills at a local college. Another option is taking short courses offered by professional societies and training firms. An increasingly popular continuing education option is to take courses offered by accredited institutions on the Internet. Many employers will partially or completely pay for education expenses that are job related.

10. Master a relevant foreign language. Business is becoming global. For instance, your company may be foreign owned, have business operations in foreign countries, or have a joint venture with a foreign company. Mastering a foreign language can be a wise career move if your employer does a great deal of business with firms in a particular country.

Appropriate language skills can also be an advantage if major technical efforts in areas of your department's interest occur in a particular foreign country. Your department will make a much better impression in meetings with foreigners if someone can speak and understand their language. Also, you are much more likely to be the one sent on trips to the foreign country. So what language should you learn? Your company's ownership, foreign business, or areas of technology probably will dictate this. Currently, there is a tremendous need for people who can speak Japanese. Germany is the world's largest exporter, so German is a reasonable choice. Latin America and China represent huge potential markets, therefore Spanish or Chinese are also reasonable choices.

Teams are also becoming more common as firms work more closely with their suppliers and customers. Long-distance teams, both within a company and intercompany, are becoming more common. Improved communications technology is promoting the use of long-distance teams. If you have not done so yet, expect to use conference calls, videoconferencing, and e-mail to communicate with long-distance team members. The more diverse a team becomes, the greater its potential to solve difficult problems and the more important individual team members' communication skills become (see chapters 3 and 4).

Companies place a high value on team players. Being a team player may not come easily or naturally. It is a good idea to look for opportunities to demonstrate that you are a team player even if you are still in school. Except for special projects, the classroom presents relatively few opportunities to demonstrate and practice these skills. Extracurricular activities (not only sports) offer many opportunities. Community service also offers a means of working together with people to accomplish joint goals. Out of school? Working chemists, engineers, and technicians can use professional society activities to develop teamwork skills. Many professional society activities, such as organizing a regional meeting and committee projects, are team efforts that develop leadership, organizational, and interpersonal skills. Civic associations, churches, and charitable groups offer many opportunities to learn and practice team skills.

RESOLVING WORKPLACE DISAGREEMENTS

Workplace disagreements. Everyone has them. The harder you work and the more you try to accomplish, the more likely your need for resources will conflict with those of your coworkers and the more likely you are to have workplace disagreements. Yet coworkers must get along with each other particularly as more work is organized and accomplished through self-directing work teams. How can you minimize the occurrence of disagreements and constructively manage them when they occur?

The process begins with an understanding of your coworkers. Most of us are reasonable sorts; logical arguments and a sense of shared mission makes it possible to handle disagreements in a civilized way. However, even the most reasonable person may have a bad day and occasionally be difficult to deal with. Some people are difficult to deal with much of the time. Often, they are highly productive or creative people. Nevertheless, difficult people's behaviors can reduce the productivity of their coworkers. Most people who exhibit such behavior do so to protect themselves from feelings of insecurity, rejection, and failure. In this era of downsizing and reengineering, difficult behavior could become more common due to increased workplace stress. Chronically difficult people fall into several types. Tactics that work to resolve disagreements with one type often will not work with another. Therefore, it is important to know their behavior. An optimistic, but constructive, way to deal with conflicts with these people is to act from the premise that there are no disagreeable people, only disagreeable behaviors. It is much easier to modify people's behavior than their personalities.

Sherman Tanks

The first type of difficult behavior to deal with is bullying. These people have been called bullies, bulldozers, and Sherman tanks. These are people who have to get their own way and do so by rolling over all opposition. In meetings, they dominate discussions and make it difficult for others to voice their opinions. Bullies often state opinions as fact. To deal with them, begin by getting their attention. Often this means interrupting them. Do this politely, but be forceful. Make eye contact and use their first name to begin a sentence. If "Sherman" continues talking, repeat his name more loudly in a determined tone without shouting. Avoid emotional or confrontational remarks such as, "You're completely wrong" or "You're out in left field." Instead, say something like, "Sherman, we need to talk more about this. I understand your position. Now let me give you mine." Suggest sitting down as tanks tend to be more confrontational if they are standing. Suppose Sherman interrupts after he has stated his position and you have not completed stating yours. Be forceful and say something like, "Sherman, I've listened to you. Please give me the same courtesy." If you can find some areas of agreement with Sherman, the remaining areas of disagreement often become much easier to resolve. Point out the areas of agreement as early in your conversation as possible. This will tend to defuse the conflict. Many tanks like having allies. Within reasonable limits, do workplace favors for tanks. If they remember these favors, they will be less confrontational in disagreeing with you. If you must, remind Sherman of a recent favor or situation in which you solved a problem together. Comment that this shows you can work together constructively. Many tanks will quiet down amazingly when you take this approach. This is a good method to use when you need a tank's cooperation and approach him with a request.

Sammy the Sniper

Many of these tactics will not work with snipers, who are sometimes called, saboteurs. Often they care only about promoting their own careers. In meetings, they will take potshots at others by criticizing them but avoid direct confrontations. Instead they will make negative remarks to other meeting attendees. They enjoy raising issues that they know will provoke disagreements among others and then sit back and watch the conflict as the meeting turns into a free-for-all. Sometimes, they will promise cooperation in front of others and then withhold it. Dealing with Sammy the Sniper can be frustrating. You need to hold him to agreements by documenting them. During meetings, draw out Sammy's disagreement so all attendees can consider his opinion and not just the one person to whom Sammy addressed his negative remark. Then, get other people's opinions. Do not give into the sniper's views; if possible, offer the sniper alternatives.

When you and Sammy disagree over allocation of resources, such as instrument time, sit down privately and work out a compromise. Be forceful but do not become a Sherman tank yourself. You will find that in one-on-one situations most snipers will become much less aggressive and disagreeable. However, the sniper may still complain about you to third parties. When you come to an agreement, document it with an e-mail addressed to Sammy and copy the concerned third parties such as instrument operators or your supervisor. In this way, Sammy will find it difficult to renege on his agreement with you or unilaterally change it. Without documentation, this is a very real possibility.

Chronic Complainer

Downsizing has brought many complainers out of the woodwork. Sometimes called whiners, jeerers, or complainers by experts, these permanent cynics are convinced nothing will work well. Where you see a challenging opportunity, they see only obstacles and future failure. They are convinced every new project will fail and every company policy has a dark side. They often feel their contributions are overlooked and are frustrated in their career advancement when they see others they regard as less capable getting ahead. Some of these pessimists will try to dominate meetings while others will sit silently in meetings and voice their complaints later in informal hallway discussions with coworkers. Hallway conversations with chronic complainers are often depressing and enervating. In brainstorming sessions, most complainers try to be the fastest gun in the west when it comes to shooting down ideas.

How can you deal with a complainer? The key is to take the ammunition out of their six-shooter. Politely interrupt them to gain control over the situation. Quickly sum up the facts. Acknowledge the valid points they raise while ignoring the rest. By speaking firmly and quickly, you can keep the meeting or discussion on a productive course. If they persist, ask them to put their complaints in writing so this way everyone can give them careful consideration. Explain that, with limited time available, the meeting or discussion must move on to other issues. Chances are the other meeting attendees will agree with you. One way to make this stick is to begin discussions with them by placing a time limit on each agenda item. "Charlie, I only have ten minutes but I wanted to talk about. . . ." is a good way to open the discussion of a subject. However, do not turn into a bully yourself and dominate the conversation.

Bullying behavior can temporarily quiet a complainer. However, you have done nothing to modify their behavior. How can you accomplish this, at least in a limited way? Be sure to acknowledge the complainer's contributions, even relatively minor ones, in corporate reports, the acknowledgments section of published papers, and oral presentations and discussions. Whenever you can, find a reason to praise the complainer's contributions during meetings. This is often difficult and may leave you resenting the complainer. However, when you do this some complainers will be less likely to interrupt you later with complaints or doubts. When you discuss a course of action, find ways to give the complainer a stake in their success. This is a good tactic to avoid arguments and later "second guessing." Do your best to avoid simplistic "I'm right—you're wrong" arguments with a complainer, particularly in front of others at meetings.

Complaining is a way for people to avoid responsibility and emotionally distance themselves from failure. This behavior, while it can be moderated, is difficult to change. However, these tactics, if applied consistently, can go far in converting the complainer into an ally who thinks you are one of the few good folks around.

Exploders

Exploders are like dynamite. Even mild displeasure with a situation can cause them to blow up and they sometimes do so unexpectedly. Their behavior includes shouting, pounding a fist on the table, getting out of their chair and striding about, or looming over the sitting object of their displeasure. With the increasing emphasis on working in teams, exploders are becoming more rare. Unlike some bullies with authority who calculatedly appear to lose their temper as a means of controling a situation, exploders are truly angry.

The key to dealing with exploders is not to exhibit emotional behavior yourself, no matter how angry you become. This can be *very* difficult. When someone explodes during a one-on-one conversation, a good technique is to forcefully interject, "I'll come back when you're ready to hear what I think about this." Give them time to regain self-control. If possible, let them come to you to resume the conversation. However, often exploders explode to avoid dealing with an issue. Therefore, do not procrastinate and drop an important issue to avoid a possible emotional outburst from an exploder. If the exploder detonates during a meeting that you are chairing, make a firm request for him to allow others to voice their points of view. Begin your request by using his first name to get his attention. Speak loudly, but do not shout. Also, as for Sammy the Sniper, do not allow your suggestion to turn the meeting into an open forum in which the original agenda gets ignored. If appropriate, declare a short recess to allow the exploder to regain self-control. What if you find it difficult to gain people's attention to do this? Try an attention-getting tactic such as switching the room lights or an overhead projector on and off repeatedly. In a truly desperate situation, postpone the remainder of the meeting. Go to your supervisor and request that another person be named to replace the exploder on your team. If you are the exploder's supervisor, make it clear that you will replace him on the team if he throws another tantrum. What if you are neither the exploder's supervisor and nor the person running the meeting? In this case, wait a couple of minutes for the chairperson to take control or the exploder to run out of steam. By this time, most of the people in the room will be acutely uncomfortable. Then, inject an emotionally neutral remark such as "I think we ought to move on" or "I think we need to take a short break." Do not shout, but say this loud enough so that most of the meeting attendees can hear you. A couple of moderately loud "excuse me's" may be necessary to first gain attention. You will find you have the enthusiastic agreement of almost everyone at the meeting. Suggesting a break is a particularly good tactic. It gives the exploder time to regain self-control or the chairperson a chance to talk to the exploder. As an attendee, the break gives you the opportunity to suggest to the chairperson that if the exploder persists in his behavior, the meeting will be a waste of time. Note that it will also be an unpleasant experience making it more difficult to get good attendance for future meetings. If you feel you must make these comments, do so in a private, diplomatic way, and avoid criticizing the chairperson.

General Tactics for Resolving Disagreements

When discussing disagreements or trying to get a coworker's cooperation, certain tactics are generally useful whether or not that person falls into the disagreeable category. First, you have to get the other person's attention. Avoid becoming emotional in response to difficult behavior. If you are trying to get agreement to a course of action or participation in a project, describe the benefits of this plan to the other person. In the case of disagreements, identify areas where both of you do agree and try to build on these. Finally, you should ask a supervisor to adjudicate your differences only as a last resort. Each of these points merits additional discussion. Before beginning a discussion, you must have the other party's undivided attention. This means picking the right place and the right time for your discussion. Do not raise the subject when the person is distracted by the physical environment or another issue she is trying to resolve at the time. If this happens when you initially try and see someone, quickly arrange a mutually convenient place and time for a later discussion.

Disagreements in front of third parties usually reflect negatively on all parties to the conflict. If you have a disagreement with one person, suggest a private meeting to resolve it. Bullies, snipers, and exploders feel more secure and often behave more aggressively when on their own "turf." So try to hold your one-on-one meetings with them on a neutral site rather than in their office. This is often preferable to holding the meeting in your office. It is more appropriate for you to leave the conference room should the discussion get emotional and you become concerned about losing your temper. The one clear exception to this is when you are the supervisor. Hold the meeting in your office to put yourself in a position of greater strength and remind the difficult person of your authority. This may make the bully, sniper, and exploder feel insecure enough to moderate their behavior. The complainer usually feels very insecure (at least subconsciously), therefore it is often a good tactic to hold your discussion in their office where they feel more secure.

With the meeting place arranged and the discussion ready to begin, you next have to capture their attention. Do this by briefly explaining the benefit to them of what you are about to discuss. For example, say "Sue, if you help me with project X, together we can complete it on time. Your name will go on the report and we'll both look like heroes since the department will meet our goal of commercializing three new products this year." Having shown the other person a reason to listen, you can then describe exactly what you would like them to do. Showing benefit, particularly when asking for a major favor, is important even with very pleasant coworkers. People need a good reason to contribute a substantial amount of time and effort to help you with a problem. Sometimes, the satisfaction of helping a coworker or a work team meet goals simply is not enough. Benefits you can provide your coworker in return for a major favor include sharing credit, doing a favor for the other person, or promising to do so. If you promise to return a coworker's favor, do so in a timely manner. Otherwise, your credibility will be gravely damaged, not only with the coworker in question, but also with other coworkers as the story of your behavior spreads. Conversely, a reputation for returning favors and doing favors for others will help you gain coworkers' agreement when you ask the same of them.

If you are having a discussion with a difficult coworker and cannot resolve the situation, suggest that you each state what you want from the situation. Then look for common ground to begin to build a compromise. If their response is to stonewall, do not threaten or become emotional. Assuming you have no authority over the coworker, ask "If you do this for me, what can I do for you in return?" Turn the disagreement into a negotiation and look for ways both of you can benefit from an agreement. If you just cannot resolve the disagreement, consider asking a supervisor to do so. This should be done as a last resort since managers respond very negatively when their staff members demonstrate they cannot manage conflict. Also, it brings the manager into a situation he is certain to find unpleasant. So both parties are likely to suffer damaged reputations.

Avoid becoming emotional. This includes obvious things like not becoming angry or defensive. It also means to be aware of how the other person may interpret your language. For example, when you are trying to understand their position, it is natural and effective to ask questions. However, avoid asking questions that begin with "why." These tend to be interpreted as attacking and can elicit emotional expressions. Instead, ask questions beginning with the word "what." These questions tend to result in more fact-based less-emotional responses. Compare "Why do you think that?" with "What are the reasons for thinking that?" The second question is less likely to draw an emotional response and its answer could provide more information that will help you find areas of agreement on which to build. While you want to avoid becoming overly emotional during the disagreement or negotiation, show your satisfaction afterward. Tell the other person you are pleased at having resolved the situation and reaching a solution the two of you are happy with or at least can "live with." By helping your coworker share your sense of satisfaction and accomplishment (and maybe relief), you can hopefully resolve your next disagreement easier.

DIVERSITY

The increasing diversity of the North American workforce increases the challenges of working in teams and getting along with our coworkers. The people in our workplaces are coming from an increasingly broad range of ethnic heritage, cultural backgrounds, and lifestyles. To work together productively, people need to have an appreciation for each others' culture and ways of thinking and accept that everyone does not work together or approach problems in the same way. Studies indicate that teams whose members have diverse backgrounds are more effective in solving problems. The globalization of technology and of markets means that companies can capitalize on the diversity of their employees to more effectively do business in foreign countries. In addition, equal opportunity is the law. All of these factors plus demographic trends mean that American workplaces will become more diverse.

How can you function effectively in a diverse workplace? Many companies have instituted expensive diversity programs. I think these programs really boil down to three basic principles:

1. Don't practice the Golden Rule. Instead, treat others as they would like to be treated.
2. You have to first educate yourself. Learn about the cultural heritage of

your coworkers. Read to understand the history and customs of people from foreign countries. Understand what behaviors others might find annoying and avoid them. Learn what motivates them. It may be different from what motivates you. This can be difficult even in workplaces most consider homogeneous. For instance, at many companies, the cultural clashes between business managers and researchers were legendary before they began working together on teams and understanding each others' concerns.

Much useful behavioral information will not be found in books. To educate yourself, ask polite questions. Explain that you want to know so you can work together more effectively. Unless you are close friends, limit your questions to workplace behaviors and issues. Engage in team-building exercises to explore what motivates each team member.

3. Be polite, but honest with others. You are entitled to have your diversity respected also. For example, to pick a minor and nonemotional issue, people from some Asian backgrounds are accustomed to standing very close to others when they converse, particularly in one-on-one conversations. Many people from northern European backgrounds in particular find this uncomfortable. This situation can be amusing to watch. As one person moves slowly back from the other, the second one edges forward. Even a situation this trivial can leave at least one party annoyed at the other. One finds it frustrating that his "personal space" is being invaded. The other interprets the person's moving away from her as dislike or rejection.

This situation and each person's annoyance with the other can be easily resolved. Our first person could tell the other, "I'm sorry, but I find it uncomfortable when people I'm talking to get very close to me."

One final point, being more accepting and tolerant of others does not mean being tolerant of behaviors that are inconsiderate to you or to third parties.

WHEN YOU ARE BYPASSED FOR PROMOTION

Another situation that calls for good interpersonal skills and planning is being bypassed for promotion or a coveted assignment. How you react can determine your future career success. There are some general rules you can follow. Do not indulge your anger and frustration. Instead, use these negative feelings as fuel to make constructive changes in your career goals and strategy. Vent your feelings away from the job. Talk to your spouse, parents, or other relatives about the situation. If you have a trusted friend at another company, review your situation with her. Do not confide in coworkers. An overheard conversation or unguarded remarks from your confidant can make your feelings known to your supervisor and coworkers.

Do not try to figure out why someone else got the promotion. Speculation is unproductive and prolongs your disappointment. Take a constructive approach and determine how you can get the next promotion. Begin by asking yourself if your expectations of promotion were realistic. What are your recent accomplishments? How long have you spent in your present assignment? Do you get along with your coworkers? Do you work well on teams? Do you take a constructive and optimistic approach to solving problems? Do you inform your supervisor of your accomplishments in an appropriate and timely manner? Negative answers to any of these questions can identify areas for improvement. If you have problems in any of these areas, you will probably need to show major improvement before being considered for a promotion.

Meet with your supervisor. Make the meeting a constructive one for both of you. If your self-assessment did not identify any major problem areas, begin by saying, "I thought that I was qualified for the recent promotion, but I didn't get it. Could you explain what skills I was lacking?" An alternative is to identify a future promotion possibility and tell your supervisor you are interested in it. Ask, "What do I have to accomplish? What skills do I need to improve to qualify for a promotion in the future?" Agree with your supervisor on an action plan for self-improvement to meet her requirements for promotion. Set up a schedule to review your progress. Then put this plan into action. By using this constructive approach, you and your supervisor are working together to develop career opportunities for you. Since your supervisor is investing time and effort in you, she will be motivated to help you succeed. Remember, a good supervisor will want you to do well and get promoted. It makes her feel good and look good to her own supervisor.

WHEN YOU ARE PROMOTED

When you do get that promotion, some coworkers will have that same sense of disappointment you had when someone else was promoted. Remember how you felt and treat your coworkers with consideration and respect. Respond to the announcement of your promotion by expressing appreciation to others. Take the people who helped you earn the promotion out to lunch or invite them to a modest celebration at your home. These folks may include technicians whose hard work helped you complete projects on time, coworkers whose ideas and assistance contributed to the success of your work, the secretary who typed your reports, and the supervisor and mentors who helped you both achieve your goals and grow professionally. Technicians, as well as chemists and engineers, should do this. Given their more-modest salaries, they may want to limit their celebration to bringing a cake or doughnuts to work. When you show your appreciation to others, do so in a low-key way. Showing appreciation to coworkers without being ostentatious about it will help others overcome any resentment they feel that you, and not they, were promoted.

Mastering Your New Responsibilities

If your promotion involves a transfer or assumption of new responsibilities, talk to your new supervisor to determine the new requirements. Find out your predecessor's goals. Modify them if you need to through discussions with your new supervisor and team members.

Be aware of team political concerns. For example, you may have been transferred to another department and promoted to be team leader of a certain project. Some team members may have coveted the team leader job and were disappointed when they did not get it. Discuss this with your supervisor to find out if this was the case. This way you can better understand their motivations should they engage in unproductive behavior.

Do not change things blindly just for the sake of change and to demonstrate your authority. The reasons your predecessor had for doing things a certain way may have been valid. Before you change something, find out why your predecessor did it the way he did. Get input from team members and others reporting to you. If they strongly prefer a different way of doing things, be open-minded and consider it. Proceed carefully in making changes. To continue our previous example, suppose your project team meets once a month. However, you think more frequent team meetings are necessary. First, analyze your reasons for this change. Be sure you do not want more meetings just to provide yourself with a sense of control and security. If the reason is valid, then talk to your predecessor and find out the rationale for the meeting frequency. Then, during a team meeting, get the team members' opinion of the meeting frequency. They may agree with you, perhaps only to appear accommodating and pleasant. They may disagree with you. Should you still feel more frequent meetings are necessary, make sure the meetings are organized and run efficiently. If the meetings are productive, valuable, and timely, your team members will willingly accommodate themselves to the more frequent meeting schedule. If they are not, the team members may complain, if only behind your back.

Do not change everything at once. If you decide to change several ways your team or work group operates, give your group advance notice and gradually implement change. For example, let us return to our R&D team meeting. You have several changes you want to make:

- Switching from a monthly to a biweekly meeting schedule
- Having a team member take notes and issue meeting minutes when no one had done this before
- Having technicians, as well as chemists and engineers, attend the meetings
- Having a marketing manager attend the meetings

Making all these changes at once could disrupt a well-established team. Begin with just one or two changes. Announce your plans and schedule for making any changes you plan. Phase them in gradually on an appropriate schedule. Returning to our example, you could begin with just two changes: having the marketing manager attend and asking or assigning someone to take notes and issue a list of action items agreed upon and decisions made at the meeting.

Be willing to change course if the benefits of your changes are not what you expected or if they interfere with the team achieving its goals. Being flexible will earn you the cooperation and perhaps the admiration of your team. Let us return again to our team meeting. Your team members approve of team minutes and action items. However, the marketing manager has been missing some of the meetings and dominating the ones he has attended. He brings up matters peripheral to team activities and often causes the meetings to run overtime. To deal with this, return to the practice of not having the marketing manager attend. However, handle the situation diplomatically. First, talk to him privately and note that you realize attending team meetings is often inconvenient due to his busy schedule. To let him know what is happening without his attending the meeting, tell him you have decided to send him a copy of the minutes. If he has concerns about goals or the rate of progress in accomplishing team goals, he can discuss them with you privately.

By first expressing appreciation to others and then learning your new responsibilities and exercising your authority carefully and diplomatically, you will minimize coworkers' resentments and increase their commitment to fulfilling common goals.

OTHER WORKPLACE BEHAVIORS

Your workplace behavior can aid you in your career and make others seek you out. Alternatively, your behavior can cause you problems and make it more difficult to accomplish your goals. Few people want to associate with an individual exhibiting the various difficult behaviors discussed above. This is true even when the objectionable behavior is not aimed at you. For example, a coworker who is always complaining about company policies will soon wear out his welcome in your office. Far from seeking out his company during lunch breaks and in meetings, you will find yourself avoiding him. So if you catch yourself exhibiting one or more of these objectionable behaviors, stop! These behaviors can easily evolve from unusual occurrences to bad habits.

Changing workplace cultures can mean that behaviors once considered perfectly acceptable, even encouraged, become objectionable and must be stopped. This happened to me. When I began working for my current employer, the workplace culture was command and control. Chemists told technicians exactly what to do and how to do it. Technicians were expected to follow instructions without deviation and often without questioning them. "Thank-you's" for jobs well done were rare. Indeed, I was considered liberated for listing technicians as coauthors of external publications. Coworkers admired me for my ability to get things done as I supervised as many as six full-time technicians at once. Then my company adopted employee empowerment and a team-oriented approach to projects. Suddenly, mutual agreement on goals and deadlines became the accepted mode of working together. Chemists and technicians would agree on procedures and requirements. Then the technicians would work independently until the work was completed or the deadline arrived. For the chemist to check on progress before the deadline often required diplomacy since many technicians treasured their newfound freedom.

This approach gave chemists more time for other more challenging and enjoyable tasks. However, there were others and I who were slow to adapt. This led to disagreements with technicians and reduced productivity. Technicians used to be eager to work for me because I aggressively sought promotions and career opportunities for productive technicians (and I still do). However, this eagerness changed to reluctance as the empowerment culture became established in our workplace and my adaptation to the new workplace culture was slow. It took months for me to accept that I had a problem and even longer to modify my behavior and not lapse into old habits, particularly under stress. Even after I modified my behavior, it took a long time to dispel the reputation I had developed as a command and control freak.

YOUR PERSONALITY AND WORK STYLE

Your personality expresses itself in your everyday behavior, including your workplace behavior. Some of us exhibit behavior that is not only unproductive; it is counterproductive such as my tendency to try to overcontrol others, as discussed above. Such behavior can make you one of the "difficult people" discussed earlier. This behavior can include impatience, distrust of coworkers, an authoritarian style, not giving coworkers credit for their accomplishments, and many other behaviors. These behaviors can reduce your effectiveness and productivity even if you have excellent technical skills.

Fortunately, you can modify your behavior without taking on the much tougher assignment of modifying your personality. In today's empowered, team-oriented workplace, an authoritarian style is usually counterproductive. If you, like me, are an impatient, task-oriented person, you will have to find appropriate outlets for this trait. These outlets should not be other people unless they are clearly negligent in their duties. If they are diligent avoid pressuring them. Settle on a feasible, mutually agreeable deadline and allow your coworkers to proceed. So what should be the outlet for your impatience? You can turn it inward. Even a full-time manager can take on assignments that mean writing reports and accomplishing things on their own by a given deadline. The problems you have in meeting deadlines and accomplishing goals can be useful reminders that help you curb your impatience with others. Impatience is often a symptom of an aggressive behavior. Other forms of aggression in the workplace include seeking out disagreements with coworkers such as arguments over "turf," being unpleasant and overchallenging during meetings, and bullying coworkers in an effort to get things done. Sports are often a good way to vent aggression so it is less of a problem on the job.

However, many scientists are mild mannered and actually need to develop more decisive, forceful behavior on the job. They fear offending coworkers and often let themselves be taken advantage of. Activities that help develop self-confidence can aid in overcoming these tendencies. Job-related activities could include serving on important teams and presenting papers at conferences. Hobbies that provide a sense of achievement can also help.

Perhaps the most difficult behavior to modify is lack of trust. Many of us spent much of our professional lives in a relatively authoritarian atmosphere. This often makes it difficult to delegate authority and responsibility and trust coworkers to complete assignments properly and on time. The point of this extended discussion is that you should analyze your workplace behavior in order to identify weaknesses and then devise a strategy to overcome them. There are many possible strategies, including increased self-awareness of your behavior, counseling, and taking courses. My former employer actually offers an assertiveness training course to help individuals overcome the problems described above. (It is very popular among technicians. I think all the technicians I have worked with have completed this course!)

In chapter 2, keeping a time log as an aid to effective time management and to improve productivity was discussed. You could also keep a behavior log. Note the times and dates when you engaged in behavior that actually reduced your effectiveness. Many of us are relatively insensitive and often are unaware when we engage in such behavior. For example, I know I was often unaware of being impatient with others. The act of keeping, and occasionally reviewing, a behavior log can sensitize you to your behavioral problems. I know my behavior log helped me realize the seriousness of my problem. The behavior log can help identify situations or individuals who seem to trigger your unproductive behavior. Understanding the circumstances in which you behave inappropriately can help you understand why you do so. An understanding of these triggering situations or individuals, can aid you in minimizing your unproductive behavior.

If you are not an incorrigible example of one of the "difficult people" discussed in earlier in this chapter, coworkers may be able to help you overcome unproductive behavior. Ask them to remind you gently when you engage in the behavior you are trying to moderate. I was able to persuade my technicians, instead of getting angry when I got too impatient, to say, "John, you're doing IT again." This clear, but justifiable, reminder was enough to make me either back off when I was being unreasonable or explain the reasons for my impatience. While this strategy was helpful, perhaps the most effective technique I use is mutual agreement with coworkers on deadlines. Not being impatient does not mean being a wimp. I remain quite forceful in determining why a mutually agreed upon and logical deadline was not met.

ADDITIONAL READING

Teamwork Skills

Cain, J. and Jolliff, B. *Teamwork and Team Play.* Kendall/Hunt: Dubuque, IA, 1998.

Covey, S. R. *The 7 Habits of Highly Effective People.* Simon & Schuster: New York, 1989.

Harrington-Mackin, D. *Keeping the Team Going.* AMACOM: New York, 1996.

Larson, C. E. and Lafasto, F. M. *Teamwork: What Must Go Right, What Can Go Wrong.* Sage: London, U. K. and Thousand Oaks, CA, 1989.

Eichols, M., Bartanen, M., Narver, K., and Lewy, D. *Business Relationships: The Dynamics of Teamwork.* Partners: Madison, WI, 1997.

Parker, G. M. *Team Players and Teamwork: The New Competitive Business Strategy.* Jossey-Bass: San Francisco, CA, 1996.

Senge, P. M. and Senge, P. N. *The Fifth Discipline: The Art and Practice of the Learning Organization.* Currency/Doubleday: New York, 1994.

Resolving Workplace Disagreements/Negotiating with Coworkers

Aubuchon, N. *The Anatomy of Persuasion: How to Persuade Others to: Act on Your Ideas, Accept Your Proposals, Buy Your Products or Services, Hire You, Promote You.* AMACOM: New York, 1997.

Bramson, R. *Coping with Difficult Bosses.* Simon & Schuster: New York, 1994.

Bramson, R. M. and Bramson, R. N. *What Your Boss Doesn't Tell You Until It's Too Late: How to Correct Behavior That Is Holding You Back.* Fireside: New York, 1996.

Bramson, R. N. and Bramson, R. M. *Coping with Difficult People.* Bantam Doubleday Bell: Des Plaines, IL, 1988.

Brinkman, R. and Kirschner, R. *Dealing with People You Can't Stand.* McGraw-Hill: New York, 1994.

Cohen, A. R. and Bradford, D. L. *Influence without Authority.* Wiley: New York, 1990.

Fisher, R., Ury, W., and Patton, B. *Getting to Yes.* Penguin Putnam: New York, 1996.

Haden Elgin, S. *The Last Word on the Gentle Art of Verbal Self-Defense.* Prentice-Hall: New York, 1987.

Kindler, H. S. and Keppler, K. *Managing Disagreement Constructively: Conflict Management in Organizations.* Crisp: Menlo Park, CA, 1997.

Leritz, L. *No-fault Negotiating: A Simple and Innovative Approach to Solving Problems, Reaching Agreements, and Resolving Conflicts.* Casa Pacifica: Portland, OR, 1990.

Raiffa, H. *The Art and Science of Negotiation.* Belknap: Cambridge, MA, 1985.

Diversity

Cross, E. Y. and Blackburn White, M., eds. *The Diversity Factor: Capturing the Competitive Advantage of a Changing Workforce.* Irwin Professional: Toronto, ON, 1996.

Loden, M. *Implementing Diversity: Best Practices for Making Diversity Work in Your Organization.* Irwin Professional: Toronto, ON, 1995.

Norton, J. R. and Fox, R. E. *Change Equation: Capitalizing on Diversity for Effective Organizational Change.* American Psychological Association: Washington, DC, 1997.

Thomas, R. R. *Beyond Race and Gender: Unleashing the Power of Your Total Workforce by Managing Diversity.* AMACOM: New York, 1992.

Thomas, R. R. *Redefining Diversity.* AMACOM: New York, 1996.

General

Kelley, R. E. *How to be a Star at Work.* Times Business, New York, 1998.

PART II
TODAY'S NEW WORKING WORLD

The environment in which scientists, engineers, and technicians work has changed greatly and continues to do so. These changes are the reason for the increased importance of the professional skills discussed in Part 1.

The large-scale changes in the workplace and broad trends resulting from corporate restructuring, downsizing, globalization, and other business trends are discussed in chapter 8. In chapter 9, the focus narrows to the individual and how to succeed in this new work environment.

One response to this evolving work environment is to adopt an alternative career, which is the subject of chapter 10. Whether you enter an alternative career or remain in one of the traditional roles of scientist, engineer, or technician, the chances are good that you will be entering the job market more frequently than your peers of past decades. Consequently, maintaining your marketability in today's competive job market is an important concern. This subject is discussed in chapter 11.

8

THE REENGINEERED WORKPLACE

Corporate downsizing and reengineering have altered the nature of employment and career advancement for professionals. The dynamic nature of business is clear from the record numbers of new company start-ups and corporate mergers, takeovers, and divestitures; and this trend will continue. The only workplace certainty is change. As a result, relationships between professionals and their employers have undergone major changes. What are these changes?

Perhaps the biggest single change and one that encompasses many other changes is the loss of certainty. The rules that served experienced professionals well for decades are no longer valid. In addition, the current workplace environment is not static as companies unendingly search for methods of structuring that will improve profitability. This change extends to compensation as companies look for ways to tie individuals' compensation to the economic performance of their particular business, as well as overall company performance (see below).

The old unwritten contract between professional and employer was that continued good job performance meant continued employment and appropriate rewards in the form of promotion and raises. To a large degree, a paternal management guided the career development of professional employees. Many employers invested in the careers of their professional employees by sending them to short courses, paying tuition for evening or weekend college courses, and paying their professional society memberships. Chemists and engineers seldom changed jobs and spent one's entire professional career with one employer. All these "rules" have been revoked. Corporate downsizing and restructuring have meant that thousands of chemists and engineers have lost their jobs, with this trend continuing into at least the near future. Job loss is no longer primarily associated with poor performance. It is far more often due to changing employer priorities, the sale or closure of a business, outsourcing, and reengineering work processes to achieve greater productivity with fewer employees. As a result, most chemists and engineers can expect to work for three to six employers and to work in more than one technology area in the course of their careers.

Shorter job tenure means less of an incentive for employers to invest in the careers of their professional employees. Management delayering and downsizing mean that supervisors now have more, as well as broader, responsibilities. They often do not have the time, even if they have the inclination, to guide the careers of their staff members. With more work being done on teams, employees have less contact with their formal supervisors. This also contributes to less career guidance. It is now a mistake for professionals to expect their employer to manage careers for them. However, chemists, engineers, and technicians can adapt to this changing work environment and have rewarding and successful careers. Before discussing how to do so, let us look in more detail at how the work environment is changing.

THE HORIZONTAL CORPORATION

Companies are reengineering the way they perform many of their functions. Traditionally, corporations are organized in vertical pyramids composed of many workers with few or no management responsibilities who report to managers. Managers in turn report to more senior managers, which gives rise to many layers of management and the pyramid structure. Departments are organized around the work function. Companies are structured into departments such as sales, production, and research. Workers look upward to managers for guidance in how to do their jobs. Information flows up and down the pyramid often being filtered through layers of management. Instructions typically flow down, stopping and being interpreted by the various layers of management. All this made for slow action and inefficiency.

In today's global economy, no company can afford this inefficiency. As a result, today's trend is toward flatter organizations structured around processes such as developing and delivering new products to market. The old model had researchers, design engineers, marketers, and production staff working in separate departments in near isolation. Due to this lack of communication, problems often arose. Many times, designs of new products did not meet customers' needs. Designs were finalized with little thought about how difficult new products would be to manufacture. New product development moved slowly from one phase to the next.

In the new model, these different groups of people work together as multidisciplinary teams (chapter 7) from the start of a project. For input and guidance, team members look horizontally to other team members rather than vertically to supervisors and managers. Instead of looking inward to management for guidance, teams look outward to the customer. Information and results flow horizontally to fellow team members, as well as vertically to managers. This makes it possible to identify problems such as potential manufacturing difficulties early in a project rather than late. Teams focus on a particular goal, not on each team member's job function. Organizing by task allows companies to reduce the layers of management. Self-managing project teams are the building blocks of the horizontal organization. They conserve the resources and effort that the vertical organization devotes to managing relations among departments and sending information up and down the pyramid. Because of this, chemists, engineers, and technicians can direct more effort to satisfying customers' needs. Being on a team gives members a broader perspective of other aspects of the employers' business.

The horizontal organization increases contact between suppliers and customers. Some firms have established joint supplier-customer teams to develop new products and improve the supplier's services to the customer. The horizontal corporation also means there will be fewer promotion opportunities for professionals. Both employers and employees are rethinking the definition of career development. Their conclusion: it does not necessarily mean upward mobility. Chemists, engineers, and technicians will have to look for much of their career growth through lateral transfers that expand their capabilities and provide the stimulation of a different working environment. Of course, they will have to be alert to opportunities for promotion when they do arise. These may occur with one's current employer or require a job change.

Many corporations are shedding some functions to focus on their core profit-making functions. As employers outsource more services, an increasing number of chemists, engineers, and technicians will work for firms that provide these services. These include contract research organizations and consulting firms. As corporate needs change, so will outsourcing. These changing needs will affect the careers of many chemists, engineers, and technicians.

EMPLOYEE PERFORMANCE AND COMPENSATION

Employers are changing the way they evaluate employee performance and compensate employees. One new method of performance evaluation is "360-degree appraisals." In this process, peers and subordinates, as well as managers, evaluate an individual's performance. According to a February 1993 survey by Wyatt Corp., 26% of United States companies use 360-degree appraisals; and this percentage is increasing. Some companies such as Amoco and AT&T use these appraisals only to guide employees in how to improve their performance. Other companies such as Honeywell and Shell decide employee raises based on 360-degree appraisals. Peer and subordinate appraisals mean that interpersonal skills will become increasingly important in employee compensation. Professionals must work with peers and subordinates in a productive, nonabrasive manner and have good conflict-resolution skills. Those who try to bully others to get their way and provide poor support and guidance justifiably can expect to receive poor 360-degree appraisals.

At some companies, the manager and employee mutually agree upon a set of goals for the employee. Usually, these are annual goals. At the end of the appropriate time, the manager and employee determine how well these goals were met. Reasons for not accomplishing the goals are determined. The results of this assessment determine any changes in the employee's compensation. Some companies are beginning to tie employee performance appraisal and pay systems to team, rather than individual, performance. This encourages staffers to develop multiple skills rather than only specialized knowledge. Major employers of engineers such as Texas Instruments and Compaq do not give bonuses to individual employees; they give them to teams. The team decides how to divide the money among its members.

Many firms are tying individual compensation to company financial results. Different methods are used, but include relating individual compensation to profits, profit margins, and market share of the individual business unit and the company as a whole and benchmarking against competitors. Competitors are usually firms that are similar to the employer. Thus, a pharmaceutical firm will benchmark itself against other pharmaceutical firms. Individual business units may also benchmark themselves against competitors in individual business areas. For example, the lubricants division of a diversified chemical company may benchmark itself against a specialty chemical company whose primary business is lubricants. (This is because large companies often do not report business results for their individual divisions or business units.)

Employees whose financial compensation is tied to their employer's business success will receive fluctuating incomes as their employer experiences varying degrees of success in the marketplace. Gone is an employee's certainty of at least static, if not an ever-increasing, salary. Changes extend to other benefits also as employers offer more health care and employee savings plan options to their employees. These plans once experienced few changes with time; now changes often occur annually.

It will not be easy to adjust to these new systems—and the systems are not static—they continue to evolve. Philip Carroll, former president and CEO of Shell Oil comments, "Continuous change is the new reality." His recommendation to professionals: "Run your affairs like a business, identify your customers, develop new products, and keep improving." How can we do this?

ADAPTING TO CHANGE

Professionals can prepare themselves for this new environment, even if their employers have not announced any organizational changes. They should emphasize career advancement over the increasingly less-available promotion. Career advancement is a change in job responsibilities that broadens your skills. With fewer traditional promotions available, these lateral transfers can provide the stimulation of learning new skills while broadening your experience. It can also add to your network of corporate contacts, open additional promotion paths, and combat burnout. This broader experience can also improve your future employability.

A lateral transfer may be a long-term assignment such as a transfer from R&D to sales or marketing. It could also be a short-term duty such as a four-month stint managing a plant start-up. It is important to choose a move that will build on old skills while you master new ones. One example is a staff engineer becoming a technical representative, team leader, or patent agent. Another is a researcher becoming a marketing manager.

Narrow technical specialization is becoming less important in driving career advancement. For example, R&D chemists and engineers need to become more knowledgeable in related areas such as patent law. Sales personnel and marketers need to become familiar with the basic technology of the products and processes they sell. All employees must understand more about their customers' businesses and problems. With this understanding, they will be more effective and productive when helping their customers solve problems.

EDUCATION IS NEVER-ENDING

The need to stay current in your speciality while acquiring new skills will drive both formal and informal continuing education. Some employers will provide in-house continuing education opportunities such as short courses, and financially subsidize other forms of continuing education. There are two reasons for them to do so. First, continuing education enables employees to acquire new or improved skills that are valuable to the employer. Continuing education is a means for employers to maintain a competitive edge and improve employee productivity. Second, providing continuing education opportunities increases employee loyalty.

Of course, many firms provide little financial support for professionals to continue their education. Professionals working for these companies wisely pursue appropriate continuing education options with the attitude that they are investing in their future. Carefully chosen continuing education options can increase the value you deliver to your current employer, while also increasing the skills you can offer other firms thus improving your employability. So no matter who pays, professionals should look for opportunities to take courses that broaden their technical and organizational skills. Useful courses include time management, the basics of patent law, leadership and team participation skills, business writing, and presentation skills. Tutorials on popular computer programs can also be helpful for the novice user. Many large firms offer their own courses in these areas. Professional societies frequently offer a wide variety of short courses outside traditional technical areas to enable their members to improve their teamwork, communications, and management skills. As continuing education becomes ever-more important to career success for their members, these course offerings will increase. An increasing number of colleges, universities, and commercial organizations offer short courses and evening and weekend classes. University courses available on the Internet is a new phenomenon ideal for frequent business travelers who would often miss sessions of evening classes. Video- and audiotape courses are available for loan at many companies, local sections of engineering societies, and some public libraries.

Researchers should consider attending seminars in their field at local universities. Not only can this inform you of new developments in your own and related fields, it also can help you establish a network of professional contacts. Local meetings of science and engineering societies and other professional organizations provide the same benefits.

PROFESSIONAL ORGANIZATIONS

Professional societies provide other opportunities beside those of continuing education and developing teamwork skills. Many now provide seminars and workshops on job-hunting skills such as resume writing and interviewing. The Internet is providing new opportunities for professional societies to serve members in a cost-effective way. For example, some professional societies including the American Chemical Society maintain on-line databanks on which members can place resumes and companies can search for employment candidates. Members can often place situation-wanted ads in engineering society publications at reduced rates.

Roy Koerner, 1995 President of the Society of Petroleum Engineers, noted, "As companies begin to move and evolve, they are going to provide fewer employee programs than we've come to expect." He believes professional societies will fill that vacuum. There are already examples of this. Many professional societies offer members, long-term disability and health and life insurance. Unlike employers' benefits programs, these will remain in force when you change jobs.

As a young scientist or engineer, what organizations should you join? One should be the major professional society in your specialty. For example, a mechanical engineer could join the American Society of Mechanical Engineers. A petroleum engineer should consider joining, the Society of Petroleum Engineers. You may wish to join more than one professional society. One organization cannot fill all of your professional needs. So, for example, some American Chemical Society members belong to other groups such as the Catalysis Society, the American Institute of Chemists, and chemical societies aimed at members of various minority groups.

You should not limit your professional activities to only discipline-oriented professional societies. Many industries have trade associations. These provide opportunities to network with professionals in other fields who work for your employer's suppliers, customers, and competitors. For example, as a chemist who develops products for use in the paper industry, I belong to the Technical Association of the Pulp and Paper Industry and the Canadian Pulp and Paper Association, as well as the American Chemical Society. A cosmetics chemist may well belong to the Chemical Specialties Manufacturers Association. You can learn from people whose perspectives are much different that yours. This can be a valuable experience. For example, both engineers and sales people are active in trade associations. They work together on technical committees and many present technical papers at trade association conferences.

Employers are more likely to pay membership dues to trade associations than to scientific societies. This is because the short-term business benefits that employee membership provides to employers are clear. These include presenting papers that describe new products or processes to potential customers in a noncommercial setting. Trade associations provide an arena for professional activities that result in potential customers developing favorable opinions of supplier's personnel and their employers. They also provide an arena for solving industry-wide problems and sharing solutions. Employees can network with suppliers, current customers, and potential customers developing relationships that can lead to new business opportunities.

NETWORKING AND MENTORS

Professional society and trade association meetings and activities also provide excellent opportunities to make new networking contacts and maintain old ones (see chapter 3). By offering perspectives different than your own and those of coworkers, networking contacts can provide useful information and advice on solving problems. These include technical problems, interpersonal problems at work, and what to do in various career situations. Ask your contacts to explain how they work and how their organizations function. By doing this, your network members can be useful sources of ideas on how you can work more effectively. In addition, they can be useful sources of information both on the job market and in employment opportunities at other companies. Some can provide advice on major career changes such as leaving an engineering staff position to take a marketing assignment.

To a large degree, networking contacts can substitute for mentors. Mentors have always been important in career development, particularly for young professionals beginning their careers. However, in today's business environment, it is harder to find a mentor as people seem busier than ever. As a result, many professionals will not have a single mentor with whom they can spend a great deal of time. All professionals (and younger ones in particular) should try to develop three types of mentors. One should be a supervisor, team leader, or experienced coworker who can provide technical advice. Another should be a coworker skilled in the business areas of your company. This mentor can be particularly helpful if you are considering a major career shift from technical work to management, marketing, or sales. The third type of mentor should be someone outside your company who is knowledgeable about the industry or technical field in which you work. He can give you a different perspective on technical, business, and career issues than you can get from coworkers. You can also use your outside mentor as an information source for employment opportunities.

The more good mentors you have, the better. In addition to providing different perspectives on your problems and career situations, with more mentors, you can spread out the "work load." Also you are more likely to keep that person as a mentor for a long period by not imposing on them too much. The issue of imposing too much on a busy mentor is an important one. As a result, "students" should look for opportunities to aid their mentor. This can include explaining new technology or methods to them. Beside chemical or engineering technology, a particularly helpful activity can be to help mentors improve their computer skills. Pointing out technical papers the mentor may have been too busy to read can also be helpful.

As professionals develop and mature in their career, their mentors evolve into peers and useful networking contacts. The exchange of information and advice becomes more equal. The mentor may view this as a reward for the effort put into advising the young professional. As professionals mature in their career, they should become mentors themselves. People can both have mentors and mentor others. This can begin while professionals are still students, and junior and senior undergraduates (and graduate students) advise high school students and younger college students. Post-docs often advise graduate students. Meanwhile, many of these people are being advised and mentored by their professors. Industrial chemists and engineers should do the same. Sometimes there is no better mentor for a new employee than a coworker who has recently gone through the same adjustment process the new employee is experiencing. Such mentoring is short-term, but can lead to a career-long friendship.

PRESENTATIONS AND PUBLICATIONS

Presenting papers at engineering conferences and trade association meetings (chapter 3) can be a worthwhile exercise for both you and your employer. Employers see such presentations as opportunities to highlight their problem-solving skills to an audience of prospective customers. An outstanding presentation might even become a news story in a trade magazine. Presentations provide a way for professionals to demonstrate their professionalism and competence to peers and add to their network of professional contacts. Questions and discussions after a presentation can provide speakers with new insights and ideas. The professional contacts they develop can be useful sources about career opportunities and be references for job hunters.

Research and application papers are another way for individuals and employers to prove their technical expertise to potential customers. These are often better suited to publication in industry trade magazines than in academic journals. Coauthors with different areas of expertise can learn from each other and develop new ideas for future collaboration. Coauthoring papers with colleagues who are your firm's customers can be particularly rewarding.

ADDITIONAL ACTIVITIES

Professionals should not limit their contacts with customers to trade association meetings. In the spirit of teamwork, they should occasionally accompany sales representatives on customer calls. Discussing customer problems confidentially on their own turf can lead to improved products and services. Scientists and engineers can present seminars on new technical developments and discuss how they can benefit the customers. Often, the researchers can better handle in-depth questions than can sales representatives. Frequently, there is nobody better to address important concerns such as product safety- and-handling procedures with customers' personnel than the supplier's technicians or plant personnel who work with these chemicals or equipment every day.

Meeting with customers, working in teams, making presentations, and writing engineering papers all require excellent communication skills. Rex Olsen, President of the consulting firm Executive Horizons, notes, "The only security available is in what people know and what they can communicate." Besides continuing education courses, thousands of professionals participate in Toastmasters International to improve their public-speaking skills. Some employers encourage this by providing meeting facilities for individual clubs restricted to their employees.

Other activities also offer the advantage of broadening your skills and providing the stimulation your job may lack. Today, quality improvement is high on the list of corporate goals. Reading and taking courses to become more knowledgeable in this important area and using this knowledge at work can set you apart from your peers.

Globalization is another trend that has a major impact on business. United States corporations continue to expand abroad while foreign firms increase their operations in the United States. American and foreign firms are uniting in business alliances; trade barriers continue to fall. Engineers can profit from these trends by learning a foreign language. For example, NAFTA (North American Free Trade Act) can increase career advancement, travel, and foreign assignment opportunities for Spanish-speaking engineers. If you want to do business in Quebec, speaking French is very useful. Many professionals will visit or work in a foreign country or work extensively with foreign nationals. Understanding their customs and culture will be valuable in working amicably and productively with them. Knowledge of the history of a particular country can be helpful in understanding people, providing subjects for nontechnical conversations, and enable you to learn more and better enjoy your time in a host country. Except for history, courses describing foreign cultures are less available than language courses. So the primary way to learn what you need to know is independent reading. Assess your career field and current employer to see if knowledge of a particular culture will be helpful in your career before beginning your reading. A gradual acquisition of knowledge is more productive than madly cramming just before you go on an overseas business trip.

ATTITUDE ADJUSTMENT

The many changes occurring in the workplace are irreversible. Railing against them and wishing for a return to the good old days (assuming they ever existed) is wasted effort. Professionals need to adapt to the many changes in the workplace. In particular, this means adapting to uncertainty. This includes uncertainty in the length of your current job employment with your current employer.

The career-related adaptive techniques discussed above and elsewhere in this book will help professionals adjust to this uncertain world. In addition, professionals must also manage their personal financial affairs to better prepare for possible periods of unemployment, as well as their own retirement. The increased prospect of holding several jobs and being self-employed during your career means better financial management is a necessity.

ADDITIONAL READING

Corporate Reengineering

Carr, D. K. and Johansson, H. J. *Best Practices in Reengineering: What Works and What Doesn't in the Reengineering Process.* McGraw-Hill: New York, 1995.

Hammer, M. *Beyond Reengineering: How the Process-Centered Organization Is Changing Our Work and Our Lives.* HarperCollins: New York, 1995.

Hammer, M. and Champy, J. *Reengineering the Corporation: A Manifesto for Business Revolution.* HarperCollins: New York, 1994.

Mische, M. and Bennis, W. G. *The 21st Century Organization: Reinventing Through Reengineering.* Jossey-Bass: San Francisco, CA, 1997.

Senge, P. M. and Senge, P. N. *The Fifth Discipline: The Art and Practice of the Learning Organization.* Currency/Doubleday: New York, 1994.

Globalization

Rhinesmith, S. R. *A Manager's Guide to Globalization: Six Skills for Success in a Changing World.* American Society for Training and Development: Farmingdale, NY, 1996.

Mentors and Mentoring

Bell, C. R. *Managers as Mentors: Building Partnerships for Learning*. Berrett-Koehler: San Francisco, CA, 1998.

Biehl, B. *Confidence in Finding a Mentor and Becoming One*. Broadman & Holman: Nashville, TN, 1998.

Shea, G. F. *Mentoring: How to Develop Successful Mentoring Behaviors*. Crisp: Menlo Park, CA, 1998.

The annual multivolume *Encyclopedia of Organizations* published by Gale Research, Inc., is available in many large libraries. Similar publications are available in Canada and other countries. The information provided can help you locate possible professional societies and trade associations you may wish to join. The addresses and phone numbers provided will help you get additional information on organizations of interest.

9

Your WORKDAY WORLD

In this chapter, we will discuss how to increase your workplace accomplishments, understand how your personality influences your work style, and add value to your accomplishments. In chapter 7, we discussed how to work with others, with an emphasis on resolving workplace disagreements. In this chapter, we will discuss how to ethically influence others to help you accomplish your goals, while you help your coworkers do the same. Using mentors and role models to help shape your career strategies and workplace behavior is described. How to increase the value of what you do to both yourself and your employer is also covered.

EXERTING AUTHORITY AND INFLUENCE

Your Spheres of Control and Influence

On the job, everyone has a sphere of influence. These are the activities you can control or influence. You do not have to be a manager to constructively influence events and determine a course of action. The respect in which a manager or work team holds a technical expert can result in the expert having a major influence on decision making. Of course, consultants must have both the relevant expertise and influence with decision makers; otherwise it makes no sense to hire the consultant.

Think of your sphere of influence as two concentric spheres. The inner, smaller sphere is the activities and events whose outcome you control. For ex-

---

◆ The Currencies of Influence

In their book *Influence Without Authority* (see Additional Reading at the conclusion of this chapter), A. R. Cohen and D. L. Bradford define several currencies of influence—what you can use to gain the cooperation of others. Modified to be tailored to scientists, engineers, and technicians, the currencies are:

- Idealism—aiding accomplishment of a task that has major benefits for the employer, customers, the advancement of knowledge, or the country; aiding a junior coworker in achieving success (mentoring)
- Ego-related currencies—learning new skills, receiving recognition and visibility, a sense of being an insider, friendship, understanding, emotional support, gratitude (of the person making the request), a sense of participation in important work
- Trading currencies of resources, assistance and support on other tasks, and access to information (both technical and relating to the employer) and to members of your network

Before requesting a coworker's aid, it is useful to review this list and see what you can offer in return. While this may sound manipulative, it aids both you and the other person in fulfilling your job responsibilities and achieving your goals. These currencies are often used in professional society and trade association activities such as programming and governance. Of course, whatever you promise in return, you must provide. Otherwise, you will quickly exhaust your credibility and find others unwilling to cooperate with you.

If you already have a lot of credibility, you do not have to make explicit reciprocal offers for help. Your coworkers know you will be there to help them when needed. This is particularly true if idealism is an important currency to you. In the case of fellow team members, idealism can merge with self-interest as the team works together to accomplish mutual goals.

ample, a chemist can design an experiment and have a technician follow the instructions exactly. An engineer can "ask" a secretary to format and print a set of data tables. The inner sphere of influence, the sphere of control, is relatively small for most staff scientists and engineers. And it is becoming smaller as companies encourage philosophies of empowerment. The overall result of empowerment is staff scientists and engineers, as well as managers, free themselves of tactical, detail-oriented responsibilities to focus on strategic issues.

For many, this has meant their sphere of influence, the outer concentric sphere, has expanded even as their inner sphere of control has diminished. As professionals work more on multidisciplinary teams, interact more with customers and their own coworkers who work in other specialties, and as they become more involved in trade associations and professional societies, their sphere of influence grows.

A scientist, engineer, or technician can accumulate influence by doing and reporting good work and helping others in appropriate circumstances. Hence, the necessity of good report writing and excellent presentations. These written and oral reports must be timely to have an influence on decision making. Many scientists and engineers who, in the rush of events, fail to write timely reports find themselves writing post-mortems after important decisions were made. By presenting papers at trade associations and professional society meetings, professionals can become influential in their field. Virtually without trying, employers can benefit from this influence since potential customers, impressed by an expert, are much more open to doing business with the expert's employer.

Becoming influential increases your value to your employer. Hence, wise professionals judiciously accept requests to serve on interdepartmental teams. With good service on these teams, professionals can accumulate influence that can later be used to fulfill their own responsibilities or make a career move such as a transfer to another department. An excellent way for a researcher to make a move to a sales or marketing function is to serve on a product team with markets and sales representatives. By making suggestions and assisting these business professionals as needed (without doing their jobs for them), the professional can acquire credibility and influence that will facilitate a later career move to a business function.

Participating in trade association and professional society activities is also a good way to accumulate influence that can aid your career and improve your on-the-job performance. However, the wise professionals are careful not to overcommit themselves and take on more responsibility than they can fulfill. Also, it is important to do an excellent job at what you do undertake. Doing merely an adequate job seldom helps you gain influence.

Your Sphere of Concern

Your sphere of concern is a third, larger sphere that engulfs your other spheres of control and influence. These are things that can affect you directly or that you care about deeply. However, your ability to influence these things, at least alone, may be limited or nonexistent. For example, suppose you work for a small, privately owned company. The owner announces the sale of the company to a large, multinational corporation. There is nothing you can do to alter this event, although it and its repercussions obviously are in your sphere of concern. You feel helpless; to attempt to do anything is fruitless. To give way to frustration and anger will make you lose focus and waste energy. In this ultra-clear example, there is little most of us can do. Acting alone, we have little scope for altering the event. (However, a well-organized and financed offer from a group of employees to buy the company could negate the sale if it has not been finalized.)

Many events are less clear. Professionals should carefully analyze them to determine answers to the following questions:

- Do I really want to alter the situation? Am I mistaken? Does the situation affect me as much as I think it does?

- Are there ways to profit from the situation that I have not thought of?
- Acting alone, can I alter events in a desirable way? Is it worth my effort?
- If acting alone is fruitless, can I act in cooperation with others to alter the situation?

When you take on responsibilities in your sphere of concern, but outside your spheres of influence and control, you have a high risk of failure. You can lose focus and divert your efforts from things you can influence to those you cannot. Certainly, you can develop feelings of anger and frustration.

Many professionals have wasted time and effort trying to alter situations they cannot influence. Understanding when it is profitable to spend time and effort influencing events and when it is not is an important career-success skill. Knowing when to cooperate with others to influence events you cannot influence alone is another. Indeed, the power of collective action is a primary reason for the existence of trade associations, professional societies, and unions.

It is also important to realize that people or situations in your spheres of influence and control can move out of them. When this cannot be altered, you must adapt to this situation. For example, many of the frustrations of raising a child, particularly those associated with adolescents, are due to the children maturing out of parents' zone of control and parents' diminished influence on them. Similar situations happen in the professional world as a professor's students graduate and young professionals learn, become more mature, and rely less on mentors.

One potential solution to the situation is to expand your sphere of influence. Consider the bench scientist who needs a set of samples to be analyzed in order to proceed on a project. However, the analytical chemist tells her that due to his work schedule, the analyses will not be completed for three weeks. The analyst is very busy and works in a different department. This is a situation that concerns our bench scientist, but is not in her sphere of influence or control. How can she expand her sphere of influence? One way might be for her or her supervisor to approach the analyst and explain the importance of the sample. Another is to approach the analyst's supervisor and convince him of the importance of the samples. However, the analyst could resent this approach.

A third, and better, approach is to convince the analyst that he has a stake in the project. If the analyses truly are critical, it may be appropriate for the analyst to be a coauthor on the project report. Our bench scientist could volunteer to help the analyst write his own report on the analyses. In either case, our bench scientist, by offering the analyst a stake in the timely completion of the analyses, has expanded her sphere of influence in a collegial, nonconfrontational way.

A fourth approach might be for the bench scientist herself to learn how to do the needed analyses. This moves the situation into her zone of control. Our bench scientist needs to consider the relative feasibility of both the third and fourth approaches, and whether the independence offered by the fourth approach justifies the additional work she will have to do. Usually, the more a particular type of analysis needs to be performed, the greater the benefit of performing the analysis yourself.

This extension of your sphere of influence need not and should not be done in a mindset of exploitation. This is illustrated by the above example. The third approach, enlisting the analysts' aid by giving him both a role in the project and a stake in its success, is actually a situation of mutual influence and mutual benefit. The bench scientist influences the analyst and gets her work done. The analyst gets increased credit for his work through reports thus influencing the scientist. This influence becomes clearer if the analyst enlists the bench scientists' support to purchase a new instrument or hire a technician to complete analyses in a more timely way. These beneficial professional alliances are the bedrock of teamwork and aid professionals' careers and increase their accomplishments.

YOUR PERSONALITY AND WORK STYLE

Your personality expresses itself in your behavior including your everyday workplace behavior. Some of us exhibit behavior that is not only unproductive; it is counter-productive. Such behavior can make you one of the "difficult people" discussed in chapter 7. Review the section "Your Personality and Work Style" at the end of chapter 7. Then assess your own workplace, perhaps using a behavior log as suggested in chapter 7. Your behavior log can help you identify and eliminate unproductive or counter-productive behaviors that reduce your productivity and effectiveness.

ACKNOWLEDGING AND LEARNING FROM MISTAKES

One of the most difficult behaviors to master is how to acknowledge and learn from mistakes. It is difficult for everyone to admit they have made a mistake. Even when they do not blame others, it is difficult to accept responsibility for failure. Yet, when you read the biographies of successful scientists and business people, you often learn that they were quick to recognize their mistakes, cut their losses, and modify their tactics. We all should try to do the same in our own careers.

Of course, one should begin a project or a work relationship with coworkers by carefully analyzing the appropriate situation to minimize the likelihood of making mistakes. However, if you do make a mistake, chances are you have created a problem or damaged a work relationship. If this occurs, it is important to admit mistakes and try to minimize the damage. For example, suppose you are a supervisor who inappropriately snapped at an engineer when both of you were having a bad day. It is important not to let the situation fester. The engineer feels resentful whereas the supervisor is annoyed with himself both for making the mistake and for not rectifying it in a timely way. This guilt can lead the supervisor to also feel annoyed at the engineer. Therefore, the best solution is for the supervisor to make a quick trip to the engineer's office to explain that he was having a bad day and apologize for venting his frustration on the engineer. This is far better handled face-to-face than by telephone. It is also better to do this on the engineer's turf.

Once you acknowledge a mistake, try to learn how to avoid making it again. This can be difficult when it means modifying behavior patterns. However, many mistakes are due to patterns of behavior. Correcting these patterns can have a beneficial effect on your career and reduce your workplace stress. When faced with a situation where you would have previously made a mistake such as being impatient or losing your temper, the awareness that you overcame your problem and handled the situation differently can be very ego satisfying.

As the above example illustrates, learning from your mistakes means learning how to avoid making similar mistakes in the future. It is worth emphasizing that when your mistakes are related to behavior, it can be very worthwhile, but very difficult. For my problems of excessive impatience with coworkers, it meant three years of effort. Suppressing my feelings of impatience is still often a matter of conscious effort. However, the benefits of my decreased stress level and better work relationships with others have been well worth the effort.

When It Is No One's Fault

Of course, sometimes plans do not succeed and no one is at fault. For example, many researchers are rightly reluctant to admit that they cannot solve a problem and accomplish project goals. Research managers are reluctant, particularly on an expensive project, to terminate it after spending large amounts of money, time, and effort trying to make a project work. Business managers are reluctant to admit that efforts to penetrate a particular market have been fruitless. Plant personnel in a facility experiencing severe start-up problems are often reluctant to admit the extent of the difficulties and accept that major design changes are needed.

So what should you do when working on a project that is experiencing major difficulties and is "going nowhere?" Take a mental step back. You may want to work on other assignments for a couple of days to try to develop some mental detachment from your concerns and frustrations. Then review your efforts to solve the project difficulties. Are there other methods you could try that have a reasonable chance of success? After developing your own list of options, consult with fellow team members and solicit their input. They also have a stake in project success. Then invite your most creative coworkers to a brainstorming session and solicit their ideas.

Prioritize these ideas. Consult with the originator of each promising idea and get their thoughts on implementing the idea and their opinions on its success. They may be more open in a one-on-one discussion with you than in a meeting with several other people present.

Test the appropriate ideas. If one works, great! Be sure to give the idea originator appropriate credit. If not, do not rush to your team leader or manager to report your lack of success. Instead, develop alternatives. For example, suppose your project was to develop a new synthesis of a key intermediate in the multistep synthesis of a drug compound. Determine:

- if there is existing technology you could license to produce this intermediate
- if you can purchase this intermediate or have it toll manufactured using your firm's raw materials
- if there is an alternative synthesis of this drug compound which does not require this intermediate
- if you can outsource synthesis of the intermediate to an organization whose skills make it possible that they could succeed in a timely fashion where you did not

Chances are these alternatives were already considered when the project was designed. However, it is reasonable to review these alternatives again to determine if any might be cost effective. After you have evaluated these alternatives and identified the most promising one, then approach your team leader or supervisor and inform her of the situation. The disappointment in your lack of success will be mitigated if you can present reasonable options to pursue.

MENTORS AND ROLE MODELS

Many of us consciously or unconsciously model our behavior after mentors and role models. Observing them in action can help you overcome unproductive tendencies in your own workplace behaviors, as well as guiding you in how to become a more productive scientist or engineer. For example, thinking about the patience (often the extreme patience) of my Ph.D. thesis advisor helps me deal with my tendency to be impatient with others. Unfortunately for students, many of these models are people with little or no exposure to the modern industrial workplace. Alternatively, they may be people whose behavior patterns were formed in a different workplace culture and are now out of date. I know that I have been appalled at some of the advice some young scientists and engineers have received from mentors and role models. It is not that these role models and mentors are not fine scientists and engineers with many accomplishments to their credit; it is that the strategies and tactics that served them well in "getting ahead" often do not function well in today's corporate work environment.

This emphasizes the importance of relevant role models to graduates who are just beginning their careers. Their role models and mentors are often professors—people who spend much of their time alone in their offices reading technical literature and writing research grant proposals. Those beginning their careers should identify senior colleagues with a strong track record of accomplishment and career advancement and who have the respect of their coworkers. (Career advancement does not necessarily mean becoming a manager. It may mean promotion on a technical career ladder and remaining a bench scientist or engineer. It may also mean achieving corporate and industry recognition for technical accomplishment.) Seek the advice of these individuals. Even if a mentoring situation does not develop, observe their behavior in meetings and other workplace activities. Use them as role models (it helps to have more than one role model). Identify the behaviors that help them accomplish their goals. How do they deal with coworkers? What do they do when they have a research idea? In what allied areas, such as patents, are they experts? Answer these questions and adopt their productive behavior patterns that suit your own personality.

Do not expect mentors and role models to always exhibit perfect behavior that you should copy; no one is perfect. One person might be an excellent role model for using time effectively, another might excel at cooperating with others, and so on. Therefore, you may want to have different role models for different skills and behaviors that are important in your work.

It is best to have role models in your own company so you can observe them closely. However, in your search for role models, do not neglect professionals who work for other companies that you encounter through business, trade association, or professional society activities. Role models are as relevant to technicians as they are to scientists and engineers. Increasingly, technicians are changing career tracks to become scientists, engineers and sales representatives and to hold marketing assignments. Get to know some of these people and decide if you want to follow in their footsteps. If you want to stay in the lab, observe other technicians to learn the secrets of their success and adopt their successful tactics.

In these days of changing workplace cultures, even midcareer scientists, engineers, and technicians can benefit from observing the behavior of good role models.

Finding Ways to Add Value

Many companies survey their customers and are pleased to find they rate the company's products as high quality and report that they are satisfied with the service they receive. Yet these same companies routinely lose some of these customers to competitors whose products are no cheaper and whose product quality is no higher. Why is this?

The primary reason is that the customer perceives the other company is providing higher value. What can this value be? How can you help your employer provide it? The answer to the first question is information. For example, consider the supplier that provides excellent, on-site instruction to customer personnel on the proper handling and use of chemicals or trains personnel in equipment operating procedures. This firm is providing added value compared to the supplier that merely provides Material Safety Data Sheets or an equipment-operating manual. Suppliers can provide customers with information relevant to their customers' current and future needs. This includes progress reports on corporate research programs to solve customer problems, new shipping procedures that minimize delays, easy-to-use product ordering procedures, and a host of other services. Supplier newsletters sent to customers can keep them informed of developments. To keep informed on new technical developments of interest to their industry, smaller firms in particular often rely heavily on their suppliers.

The second question has many answers. You can help your company add value to its products and services. Researchers can visit the customer periodically to provide R&D updates. Suppliers and customers can form joint development teams to solve problems more effectively than either could do alone. Often, supplier company researchers and sales representatives participate in trade association conferences, professional society meetings, and other events such as ASTM committee meetings that customers are not able to attend. In particular, plant personnel and people from small companies often are not able to attend many professional meetings. In many trade associations, supplier personnel do much of the committee and organizational work. Of course there are personal and professional rewards for the individuals who do this. However, their employer sees the main benefits as the personal contacts with current and prospective customers and adding value to what they provide their current customers.

◆The Conference Review

At many companies, conference attendees write meeting reports for coworkers who did not attend. These can include discussions with customers and peers, as well as summaries of presentations made at the meeting. Suitably edited, these reports can be helpful to a firm's customers who did not attend the meeting.

Several trade association journals publish summary reports of meetings with brief descriptions of a few technical papers. At least one publication, the journal *Progress in Paper Recycling* goes further. Summaries of paper recycling research presentations at a large number of conferences in the United States, Canada, and other countries are summarized. This is a valuable service since almost none of the journal subscribers are able to attend all of these conferences.

Companies that encourage researchers and others to publish useful articles in industry magazines and trade journals are providing a useful service to their customers. Researchers and technical service specialists should seek out opportunities to publish such articles. Many companies publish product brochures and case histories that describe how a product solved a customer's problem. These are designed to provide useful information to customers and prospective customers. You can help write these brochures.

Sales representatives can respond to customer requests for information or even anticipate them. They may find the answers themselves or consult with researchers and information scientists.

Consult with your marketing manager or team leader to see in what ways you might contribute to adding value to your employer's customers.

Additional Reading

Cohen, A. R. and Bradford, D. L. *Influence without Authority.* Wiley: New York, 1990.

Covey, S. R. *The 7 Habits of Highly Effective People.* Simon & Schuster: New York, 1989.

Strasser, S. and Sena, J. *Work Is Not a Four-Letter Word,* ch. 7. Business One Irwin: Homewood, IL, 1992.

Teams and Team Work

Cain, J. and Jolliff, B. *Teamwork and Team Play.* Kendall/Hunt: Dubuque, IA, 1998.

Eichols, M., Bartanen, M., Narver, K., and Lewy, D. *Business Relationships: The Dynamics of Teamwork.* Partners: Madison, WI, 1997.

Harrington-Mackin, D. *Keeping the Team Going.* AMACOM: New York, 1996.

Larson, C. E. and Lafasto, F. M. *Teamwork: What Must Go Right, What Can Go Wrong.* Sage: UK and Thousand Oaks, CA, 1989.

Parker, G. M. *Team Players and Teamwork: The New Competitive Business Strategy.* Jossey-Bass: San Francisco, CA, 1996.

Mentors and Mentoring

Biehl, B. *Confidence in Finding a Mentor and Becoming One.* Broadman & Holman: Nashville, TN, 1998.

Shea, G. F. *Mentoring: How to Develop Successful Mentoring Behaviors.* Crisp: Menlo Park, CA, 1998.

Bell, C. R. *Managers as Mentors: Building Partnerships for Learning.* Berrett-Koehler: San Francisco, CA, 1998.

10

ALTERNATIVE CAREER OPTIONS

Many chemists and engineers have recently gone through the often-stressful experience of developing an alternative career. Recent restructuring and downsizing of the chemical industry, the petroleum industry, the steel industry, and others which employ many scientists, engineers, and laboratory technicians is forcing many chemical scientists to develop alternative careers. Other reasons that experienced professionals begin alternative careers include the desire for novelty and change accompanying a midlife crisis and a personal trauma such as divorce or death of a loved one, which triggers a major reassessment of one's life and goals. With fewer R&D positions available, younger chemists and engineers often enter an alternative career without having worked in the more traditional careers.

One needs look no further than at the careers of recent American Chemical Society presidents to learn the scope of alternative careers. Helen Free has retired to become a marketing consultant for her former employer. When running for the office of ACS president-elect, Gordon Nelson stated that his midcareer change from industry to academia gave him a unique perspective on the problems faced by both types of chemists. Mary Good left academia initially for a high-level industrial management position and then a high-level subcabinet post with the federal government. She is now one of the managers of a venture capital firm. Many ACS presidents from industry had left the laboratory for management positions prior to their election.

Even voluntary changes that involve changing a field of research or a transfer within an organization can affect scientists and engineers severely. They must give up the expert status gained in a particular field to become a novice in another. If the alternative career involves a move or another major change in lifestyle, it can also disrupt the lives of the professional's family and friends. Changing employers or starting your own business can affect retirement security. Therefore, denying the need to change careers seems rational and planning an alternative career should be done in a well-planned manner and not undertaken in an emotional way.

So then, why do scientists and engineers change careers? While the cause of an involuntary career change may be obvious, the choice of an alternative career is not. For professionals who voluntarily change their careers, there may be a constellation of reasons to do so, rather than a single cause. Spending a long time specializing in a narrow field can give scientists a sense that their creative abilities are becoming dulled. They may become bored with their work. This can lead to burnout. This boredom has been referred to as "content plateauing." Content plateauing occurs when people know their jobs too well and feel there is not enough to learn or enough challenge. A second type of plateauing is "structural plateauing," the end of frequent promotions. With the downsizing experienced by many industrial employers of scientists and engineers, the number of senior positions is fewer at many firms. The elimination of entire levels of managers or "delayering" is reducing the number of promotions managers can earn. This results in structural plateauing for many scientists and engineers. A feature of all professions as people age, structural plateauing is occurring at an earlier age for many as promotions become fewer and the time between them greater.

ALTERNATIVE CAREER DATA

It is difficult to gauge the extent of alternative career development by scientists and engineers. John Ziman, author of the book *Knowing Everything About Nothing* which concerns scientists changing their research fields, has noted that reduced support for both academic and industrial research and development has made career changes for scientists increasingly necessary. However, there are few studies of alternative career involvement by scientists and engineers.

Some information relevant to alternative careers may be gained from American Chemical Society salary surveys. For instance, a recent survey indicates that more than 28% of Ph.D. biochemists are working in nonchemical fields. Other fields retain more of their practitioners. However, 9.2% of the Ph.D. polymer chemists, 8.3% of the analytical chemistry Ph.D.'s, and 14.7% of the Ph.D. environmental chemists reported they were working in nonchemical fields. No segregation by age groups or year of degree was noted.

In a 1981 survey, the National Research Council found that of 44,700 Ph.D. chemists surveyed, 73.8% were still working in chemistry. Of the 11 scientific fields surveyed, only mathematicians and biologists had a lower number of scientists working in the field of their Ph.D. degree. Popular alternative fields for Ph.D. chemists included engineering (5.9%), biology (4.3%), and medical sciences (3.5%). Only 3.7% of the chemists reported that they were working outside the science and engineering fields. For a similar survey sample of 1938–1975 Ph.D. chemistry graduates, the Council found that 24.1% of the chemists were engaged in management and administration. Consultants comprised 4.6% of the Ph.D. chemists. A total of 4.0% were involved in production, marketing, and quality control. The traditional fields of R&D (41.7%)

and teaching (21.6%) accounted for less than two-thirds of the chemists surveyed.

When only 1975–1980 Ph.D. chemistry graduates were considered, 69.6% of the 7400 chemists were working in the traditional area of R&D and 12.8% in teaching. This suggests an increased percentage of chemists entering industrial R&D as compared to the 1938–1975 group. Thus, more chemists are now entering an area in which alternative careers are more common. Management and administration occupied 8.5% of these chemists, while 4.5% were in production, marketing, and quality control. Consultants comprised only 1.0% of this relatively young group.

Job mobility within the various fields of chemistry is quite high. For example, in the 1982 ACS survey referred to above, only 32.1% of the M.S. and 51.6% of the Ph.D. organic chemists reported that they were working in organic chemistry. In an ACS presidential conference entitled "The Many Facets of Professionalism in Chemistry," Alan McClelland of DuPont noted that "someone with a good background in physical organic chemistry can very rapidly become a good polymer scientist." So it may not be surprising that the National Research Council found 12% of M.S. and nearly 16% of Ph.D. organic chemists had switched fields to polymer chemistry.

In contrast, polymer and analytical chemists showed the greatest tendency to remain in the field of their most advanced degree, with only about 25% leaving these two fields. Chemists showed the greatest tendency to leave the physical and theoretical chemistry fields; less than half of the Ph.D.'s remained in physical and theoretical chemistry. Ph.D. chemists generally showed a greater tendency to work in the field of their degree than did M.S. chemists.

Usually, alternative career development involves changes much more drastic than a mere change in research area. The transfer from the university to an industrial position, the corresponding switch from industry to academe, and the transition to management or sales are all common alternative career choices often made by chemists in early or midcareer. Creative scientists and engineers often find opportunities to utilize technical skills in areas such as investment banking and stock brokering. Younger scientists and engineers facing burnout and older ones trying to cope with early retirement often enter careers where it is the mental attitudes and analytical capabilities developed in practicing science and engineering rather than the technical knowledge itself that enables them to succeed. Many middle-aged and older chemists and engineers consider consulting, which can easily be a part-time activity for the retired chemist.

Alternative careerists can be divided into four groups: older and younger professionals who make the decision to begin an alternative career either voluntarily or involuntarily. While scientists and engineers may begin an alternative career at any age, major voluntary career changes generally take place when the chemist is in his late thirties to early forties or has reached early retirement age.

For example, younger chemists, particularly those without an advanced degree, may begin an alternative career directly upon graduating from college by joining the marketing force of a chemical manufacturer. Often a B.S. or M.S. chemist will spend several years in an R&D lab and then decide that a marketing position offers more opportunities for advancement. Often, a transfer to a customer service group will expose the bench chemist to customer concerns and enable him to establish contacts within his own company's marketing force and with customers. These contacts and the knowledge they bring can aid the chemist in the decision to pursue a marketing career and aid his transition to the sales force.

THE ACADEMIC OPTION

For some industrial scientists, the attraction of an academic career is the freedom from close supervision and sense of independence. Some people also feel a desire to teach and communicate their excitement about chemistry to a new generation of students. The desire to do basic research and make significant contributions to the bank of chemical knowledge are other reasons people give for switching to an academic career.

Twenty years ago, Dr. Andrew Kende left Lederle Laboratories for the University of Rochester. Professor Kende notes that most the above mentioned factors played a role in his move to academia. His 10-year career in industrial research had been very successful and he had reached the top of the technical ladder as Senior Research Fellow. The then 36-year-old industrial chemist had a good deal of freedom to pursue his own ideas. However, he felt his research progress was impeded by the small size of his work group. After a part-time teaching position convinced him that he was good at teaching and enjoyed it, he made his final decision to leave industry for the university.

In the past, many illustrious industrial chemists joined academia in midcareer. Concerns over aging faculties and limited replacement of academic retirees are restricting this career path. Younger chemists with less than 10 years of industry experience seem to be the most common of these career shifters. For example, Dr. John Meister made the switch from Phillips Petroleum's central lab in Bartlesville, Oklahoma, after less than 10 years to Southern Methodist University in Dallas, Texas, and then to the University of Detroit in the 1980s. A polymer chemist, Dr. Meister left industry to obtain the freedom to work on problems of his own choice. He is now on the faculty of New Mexico Technical Institute.

More recently, Dr. William Rixey recently left Shell Development Company after less than six years for a faculty position at the University of Houston's Department of Civil Engineering. Rixey is representative of many recent career shifters from industry to academia in that he worked less than five years in industry. He also is unusual in that, having worked in the area of soil remediation at both Shell and the University of Houston, his appointment is with the civil engineering department.

More senior engineers and chemists still occasionally leave industry for academic positions. Dr. George Hirasaki left a research position at Shell Development Company to join the chemical engineering department of Rice University while coworker Jimmie Lawson joined the research staff of the same department.

The traditional academic advantage of choosing the research problems on which one works may be less pronounced in this era of tighter research funding. Dedicated academic researchers can aspire to emeritus status, which is not an option in industrial research careers. A traditional disadvantage, lower academic salaries, may be less than thought as academicians have the option of earning additional money by consulting and teaching courses during the summer term.

THE INDUSTRIAL OPTION

Many chemistry and engineering faculty members are forced to make the switch to industry when they do not obtain tenure. Others become dissatisfied with the low salaries of junior faculty members. There are also nonfinancial motives for the switch from academic chemistry to industrial R&D. Consultant Malcolm "Mack" Pruitt has noted that "seeing one's research results used is still one of the best motivators of all for the scientist." The possibility of manufacturing processes or developing a new product based on one's research can be a significant motivating factor for both the chemistry faculty member who decides to join industry and for the academic who engages in consulting work.

Outstanding academic researchers may be hired at senior staff or management levels to revitalize research organizations. Such was the situation for Dr. Mary Good who left the University of New Orleans chemistry faculty to become a vice president at Allied Signal Corporation.

CHANGING JOB ASSIGNMENTS

Alternative careers need not involve a change of employers. Dr. Bart Bower exemplifies those making career changes while continuing to work for a single employer. In the 1970s, he was an organometallic chemist at Hercules, Inc. He then became an information scientist for much of the 1980s. Recently, he returned to the lab and is now working with chemical products for paper industry applications. However, his new assignment requires much greater customer interaction than his former bench research position. Dr. Bower's career also illustrates another trend. More chemists and engineers are making multiple shifts in their careers. I have shifted from organic synthesis to polymer synthesis and applications, and to surfactant synthesis and applications during my industrial career. Each shift was accompanied by a change of employer.

Another example is Dr. Judith Zweig who became a polymer chemist and then an R&D manager after her employment by Shell Development Company in Houston. After several years in R&D, she relocated to Shell Chemical Company corporate headquarters as a marketing representative for a new product. She had been aiming for a move to a business center for some time. However, after less than a year in her new position, Dr. Zweig resigned because her husband had accepted a position on the east coast. She accepted a position as group leader at Olin Corporation's research center in Connecticut and returned to R&D management. She eventually joined a consulting firm specializing in writing technical literature.

The shift from the lab to research management is a common one made by many chemists. Common fears are that the manager gradually becomes technically out of date. The experience of some R&D managers such as Bill Young at Halliburton Services shows that older chemists returning to the lab can still be highly productive researchers.

THE MANAGEMENT OPTION

Financial rewards play a major role in many chemists becoming managers. However, James Burke of Rohm & Haas noted that since 1982 the ranks of lower level management have been reduced drastically at many companies. Restructuring is still occurring at many firms. The greater job security that once accompanied a switch from the lab bench to management has largely disappeared. Career plateauing in management at a relatively early age is also becoming more common as corporations reorganize. These factors all reduce the attractiveness of management in large companies as an alternative career option. However, leveraged buyouts and small company start-ups provide management opportunities for chemists. All employees became stockholders when entrepreneur Gordon Cain assembled Cain Chemical Company by purchasing petrochemical operations from major chemical companies. Many chemist/managers became millionaires with the recent sale of the company to Occidental Petroleum. The average sum paid to an employee was $400,000 with the minimum payment being in excess of $100,000.

THE ENTREPRENEUR

The key to starting a successful small company is identifying an unfilled need and satisfying it sufficiently well that one's new company survives and prospers. Alfred Bader's founding of Aldrich Chemical Company is a classic story of this kind of success. Dr. Bader identified a need for small quantities of high purity organic chemicals by academic and industrial research labs. Living by its motto of "chemists helping chemists in research and industry," Aldrich has remained a success for many years.

More recently, Oklahoma chemist Glen Penny identified a need by chemical companies and independent oil companies for laboratory evaluation of commercial products and treatment fluid evaluation for specific oil-field applications. Dr. Glen Penny was less than 40 years old when he left a bench chemist position with Halliburton Services, a major oil-field services firm, to start his own company, Stim-Labs, Inc. in early 1985. Timing was critical to his success. Oil industry downsizing resulted in a need to outsource many R&D activities, and Penny's firm supplied these needs. Penny built on the experience he gained in his industrial employment. He planned ahead, equipped his laboratory, and obtained his first client before quitting his job. Since then, former Halliburton Services coworkers have joined him. He has also hired other employees and opened up a branch in England. More recently, he sold his business to Core Laboratories, Inc. He now works as a manager for this firm.

However, statistics indicate that the majority of new businesses fail and not all chemist-entrepreneurs have had an easy time of it. David Burnett founded a laboratory services business, Burnett Research Associates, in 1982 when he was less than 40 years old. He planned well and built on his previous career as an enhanced oil-recovery specialist with Core Laboratories, Inc., a firm that provided a broad range of scientific services to the oil industry. Burnett's firm had a narrow focus on the evaluation of enhanced oil-recovery (EOR) chemicals and development of EOR processes for specific oil fields. The decline of interest in EOR led to the demise of his firm.

LEAVING THE LAB

Chemist Connie Merrill gradually became a liaison between the research laboratory and business development groups at Shell Development and Shell Chemical companies. This process was aided by her active involvement in trade associations. Her own interests and drive resulted in this informal position slowly becoming more formal and recognized within the organization. This was a gradual process that took more than four years and resulted in Dr. Merrill's 1988 transfer to a business development department with Shell's corporate headquarters. While gradual, her transition enabled her to maintain her contacts and credibility in the R&D organization while she learned the intricacies of business development. Upon assuming her new position, she had instant credibility, a track record of relevant accomplishments, and a network of useful contacts within her own company and among potential customers and suppliers. After holding other positions, she became a product manager in Shell's resins business for several years. Major corporate restructuring in 1997 resulted in her transfer from Houston to London and she became manager of diversity.

Some chemists and chemical engineers become patent attorneys. They usually begin their legal education shortly after obtaining their undergraduate degree. An exception to this is chemical engineer Denise Wolfs who spent a couple of years as a chemical plant engineer before starting law school. Another exception is Mark Dickson who took a leave of absence from Shell Development Company to be the 1985 ACS Congressional Science Fellow. While working in Washington, Dr. Dickson took law courses at Georgetown University. After a year, he returned to Shell to become a patent agent while continuing to take law courses.

Some chemists leave the lab to become information scientists. The example of Bart Bower was already mentioned. Carlene Eckroade earned the B.S. degree in chemistry while working as a chemical laboratory technician at Hercules, Inc. She then became an information scientist at W.L. Gore & Associates. Many chemists and engineers who leave the lab remain with the same employer. This can be a real advantage if the professional comes to regret her decision to leave the lab and decides to return to the bench. With a proven R&D track record and laboratory contacts in place, this change often can be made without the necessity of looking for a new employer.

Corporate restructuring, mergers, and divestitures in industries that employ large numbers of chemical scientists have resulted in many midcareer chemists and engineers finding themselves suddenly working for a new employer in a new type of job. Joint ventures are a particularly unstable situation since their average lifetime is only two years. (Joint ventures are usually terminated by one partner's purchase of the other's interest. A recent example is DuPont's purchase of Merck's 50% interest in DuPont Merck Pharmaceuticals.) Adaptability is a key requirement for success in these situations. For example, when Hercules spun off part of its water-soluble polymers business into a joint venture with Henkel called Aqualon, research manager Carl Lukach suddenly found himself in a much different career. After many years as a researcher and a research manager at the Hercules Research Center in Wilmington, Delaware, Dr. Lukach found himself in Houston as the manager of Aqualon's technical services laboratory for the oil industry. With the sale of this business, he moved back to Wilmington after two years in Houston.

THE CONSULTING OPTION

Older chemists are often more likely to find themselves forced into considering alternative careers. Early retirement has become widespread in the chemical and oil industries. While older chemists and engineers face more age discrimination in the job market, they usually are more able to develop a rewarding consulting career than their younger counterparts. Of course, for older professionals, success is dependent upon their previous records of technical accomplishment, ability to interact with people, a good sense of business, and an aggressive "go-getter" attitude.

The advent of microcomputers and small, relatively inexpensive home office equipment such as telephone answering machines, photocopiers, and fax machines have enabled many consultants to reduce expenses by operating out of their homes and not needing a secretary. The microcomputer and a modem put the world of information provided by on-line databases at the consultant's fingertips. This reduces the need to collect expensive, but seldom-used, reference materials or make frequent trips to the library.

Chemists and engineers who decide to leave their jobs and become consultants have an advantage in that they can gradually accumulate the office equipment and contacts needed to make their consulting businesses a success. By engaging in high profile corporate activities, presenting and publishing papers, and becoming more active in professional societies and trade associations, professionals who organize their transitions to consulting can establish name recognition with potential clients and thus give the new business a fast start. The retiree who decides on short notice to try consulting has none of these advantages. Such was the case for many engineers and chemists who became consultants in the downsizing and restructuring of the oil-field services industry that began in 1982, and parallel developments in the petroleum industry that began in 1986. A combination of low demand for their services by the petroleum industry and lack of preparation to become consultants resulted in the failure of many of these consultants. Some of the most successful were those who transferred their skills to soil remediation technology where much of the petroleum engineering and some of the chemistry developed by the oil industry was applicable.

Employment and technology trends of the last several years have increased the opportunities for chemists to establish successful consulting businesses and other enterprises. Reduced staff mean many companies are more dependent on outside sources for ideas and problem solving skills. Companies now outsource many other services they formerly performed for themselves in-house. At the same time, corporate downsizing has increased the number of experienced chemists who are unemployed or retired and wish to continue their careers. Despite federal regulations against the practice, many experienced chemists experience age discrimination in the job market. This and the relatively small number of positions available increase the lure of consulting as a career option.

Despite increased opportunities for consultants, the decision to become a consultant should be made only after careful self-assessment. A successful consulting business requires all the skills that any other business needs. Technical expertise, while necessary, is only one of the requirements for success. As a consultant, you will have to sell results, not expertise, to your clients. Howard Shenson and Ted Nicholas, authors of *The Complete Guide to Consulting Suc-*

cess, note "Even when potential clients are aware of your talents and skills, they don't necessarily see how you can be of service to them." So the successful consultant will have marketing, as well as technical skills. You will need business and project management skills, as well as sales and marketing skills.

Acquiring these skills may leave little time to be one of the leading authorities in your technical field. However, few successful consultants are the leading authorities in their fields. Instead, Shenson and Nicholas note, "they're active, practical, energetic people who put the theory to work and make it pay."

Too many consultants concentrate on what type of work they can perform, rather than looking at the marketplace. Skilled marketers determine what skills and knowledge are needed and find a match for their own talents. Success requires that you make yourself indispensable to paying clients who see the need for your services. Typical employees each have only one client, their employer. A successful consultant usually has at least several clients. So the loss of one, while serious, is not devastating. Obviously, this is not the case for chemists and engineers who work for a single employer.

BEGINNING AN ALTERNATIVE CAREER

When thinking about beginning an alternative career, professionals first must clearly determine why they are considering this question. Doing so will help them define whether their current dissatisfactions and future ambitions can best be satisfied through pursuing an alternative career. Because of the potential problems involved, developing an alternative career should be a last resort and should only begin after careful thought.

Before taking any concrete steps, a scientist or engineer should decide which alternative careers are reasonable options for him. Discuss these with trusted associates to see whether someone with a more detached viewpoint agrees with your assessment. Review what skills are required for success in a particular alternative career with someone already successful in that field. Professional society contacts are useful in finding qualified people who can give you good advice. Do not rely on a single opinion. Talk to at least three people who are presently engaged in the alternative career you are considering. This will enable you to assemble a list of requirements for success in an alternative career. Comparison of this list of requirements with one's own strengths, weaknesses, skills, and accomplishments will determine whether there is a good fit.

Another way to collect information about an alternative career is to join the appropriate professional society. Ideally, this organization has an active chapter in your area. By talking to members at local meetings, reading the organization's magazines and other journals covering the alternative career field, you can greatly expand your knowledge and assess the suitability of the alternative career for you.

If you can, work part-time or temporarily in the alternative career field before committing to it. Some employers will allow you to do this. An example is Dr. Len Haberman, a researcher with Shell Chemical Company. A bench chemist, he had transferred to a marketing position at Shell for one year to determine if he liked this type of work more than R&D.

The more knowledge and skills one can transfer to their alternative career, the greater the likelihood of success. These may be general skills such as training in research, or specific knowledge such as being an expert in one aspect of engineering or technology. Certain aspects of business knowledge or management skills can obviously also be helpful. The more contacts one has to help in considering an alternate career, the better. This is particularly true if you are starting your own business or becoming a consultant. Certainly, the more skills and knowledge one can transfer to an alternative career, the higher the fees or salary one can earn.

Of particular importance is whether the professional is good at dealing with people; most alternative careers involve far more interpersonal interactions than conventional chemistry and engineering careers. Other important personal attributes are flexibility and the ability to think quickly on your feet and make rapid, but informed, decisions. These certainly are requirements needed to excel in teaching or consulting. Computer skills are critical to success in some alternative careers such as chemical information.

Once an alternative career field is chosen, the chemist or engineer must then look for work. The job hunter should consider companies outside those industries that traditionally employ chemists and engineers. For example, chemists have utilized their technical skills as stockbrokers and bankers to analyze investment opportunities in the chemical industry and biotechnology. Chemists and engineers will find it valuable to become a member of technical societies and trade associations in their new areas of interest. By attending meetings and reading professional news magazines and trade association journals, they can learn about the alternative profession and employment opportunities before taking any concrete steps.

The job hunter or budding consultant will have to prepare a resume or a prospectus. When doing so, professionals should emphasize transferable skills, rather than highlighting a list of positions held that may be of little interest to a prospective employer or client. One might organize the resume by skill areas such as fields of chemistry, laboratory techniques, communications, or management, and so on. Within each area, one should list relevant accomplishments. Only at the end of the resume should applicants provide a chronological list of the positions previously held. This organization will stress the job hunter's relevant skills and interest in using these skills in an alternative career. This approach could prevent the reader from losing interest when studying the applicant's employment history.

The resume cover letter should also emphasize the applicant's transferable skills. When interviewing, the applicant should stress relevant skills and should be ready to counter questions or objections of the variety: "But you don't have any experience." Previous membership in an appropriate professional society or trade association will, at the very least, enable the job hunter to both understand and use the jargon of the new field she wants to enter. Judicious use of this jargon in a cover letter can indicate to the reader that you may be a novice, but you already know the basics.

When applying for an advertised position, do not be intimidated if you lack experience. Employers always look for the ideal candidate, but usually have to settle for someone less qualified. The aggressive committed individual can rapidly become seasoned in the new position. Look for opportunities to convince potential employers of this. There is no substitute for prior experience. Consider how you can gain experience while still employed in your present position. For example, someone interested in marketing could get involved in this end of the business when a new product she has developed is being commercialized. Consider attending a trade association where your company is operating an exhibit booth. Help your marketing personnel to staff the booth and answer questions. Offer to call on customers with salesmen to begin to learn this side of the business.

If you are interested in the strategic side of the business and business development, you might get involved with technical questions that involve decisions such as choosing a toll manufacturer, determining the method of manufacture, working with customers to set product specifications, and so on. In addition to developing a basic expertise in an alternative career area, you could be setting the stage for an internal job transfer. Beginning an alternative career without having to change employers can be a major advantage.

It may take some time, but becoming involved in professional society activities can be an excellent way to develop alternative career skills. You will develop "people skills" as you work with professionals who have different backgrounds. Opportunities to develop management skills abound in professional society activities. Opportunities to develop finance and public relations skills also are occasionally available.

Another way to develop expertise in alternative career fields is temporary or contract work. This may be difficult because employers of contract workers want instant expertise in their short-term workers. However, by working in a technician or an apprentice-type position, a chemist or engineer may be able to overcome this problem. A true commitment to the career change is needed if this approach is to be acceptable. There are many frustrations and problems involved in beginning an alternative career. However, the chemists and engineers that overcome them go on to succeed and rarely to return to their earlier careers.

PARALLEL CAREERS

One of the most interesting career developments of the past several years is the growth of the phenomenon of parallel careers among scientists and engineers. The less-sophisticated name for parallel careers is "second job." However, this second term is not completely accurate as many professionals parallel their primary career with a second one in which they work for themselves. One of the obvious options for a parallel career is consulting. One of the startling conclusions of the American Chemical Society 1995 Comprehensive Salary and Employment Survey of its members was that 27,000 of its approximately 150,000 members had spent some of their time in consulting work. More recent ACS surveys confirmed that many ACS members consulted part time.

Many professors have long been consulting. However, the phenomenon has spread far beyond the academic world and includes more career options than consulting. There have always been some professionals that have spent some of their own time in parallel careers. However, the phenomenon appears to have grown dramatically in the last 10 years and continues to do so. In my own informal survey of more than 100 professionals with parallel careers, decreased job security, the desire for a greater sense of control over their lives, and a desire for a second income are the primary reasons given for beginning a parallel career.

There are almost as many types of parallel careers as there are professionals who practice them. My own field of chemistry is an example. Geoffrey Dolbear began his consulting career in 1987 while working full-time for UNOCAL. He consulted on nights and weekends and carefully avoided conflicts of interest with his employer. By 1989, Dr. Dolbear's business had grown so much that he left UNOCAL to pursue consulting full-time.

I am not the only chemist who pursues writing as a parallel career. Several of my coworkers are doing so also. We, and many others, are in good company. For example, Nobel laureate chemist Roald Hoffman has had his poetry published. Chemist Carl Djerassi has become a successful novelist.

Some scientists and engineers look for activities that allow them more contact with people or to satisfy their other interests. While working as a chemist at Halliburton Services, Clay Cole has sold real estate while working full-time. Shell chemist Greg York does framing work, specializing in sports memorabilia. He has advertised in national magazines and he and his wife have exhibited at trade shows. Physicist Cathy Flick was teaching physics at Earlham College when she began using her Russian language skills to do translating on a free-lance basis. Eventually, Dr. Flick became a full-time translator. While teaching at the University of Arkansas, chemist Neil Ostland utilized his computer skills to develop molecular modeling software. He developed the widely used Hyper-chem® molecular modeling software. He subsequently founded and became

CEO of Hypercube, Inc. The cases of Geoffrey Dolbear, Cathy Fick, and Neil Ostland illustrate that parallel careers can eventually become a full-time job and primary source of income.

Additional Reading

Cook, M. F. *Consulting on the Side: How to Start a Part-Time Consulting Business While Still Working.* Wiley: New York, 1996.

Helfand, D. P. *Career Change: Everything You Need to Meet New Challenges and Take Control of Your Career.* Vgm Horizons: Lincolnwood, IL, 1995.

Holtz, H. *How to Succeed as an Independent Consultant,* 3rd edition. Wiley: New York, 1993.

Holtz, H. *The Business Plan Guide for Independent Consultants.* Wiley: New York, 1994.

Owens, F., Uhler, R., and Marasco, C. *Careers for Chemists: A World Outside the Lab.* American Chemical Society: Washington, DC, 1997.

Rodmann, D., Bly, D. D., Owens, F., and Anderson, A-C. *Career Transitions for Chemists.* American Chemical Society: Washington, DC, 1995.

Winter, J. *Making a Living Without a Job: Winning Ways for Creating Work That You Love.* Bantam Doubleday Dell: New York, 1996.

11

Maintaining Marketability

With professionals changing employers more frequently than in the past, maintaining your marketability, the ability to get a job, is critical to career success. Maintaining marketability means keeping up-to-date in your field and in relevant professional skills, and documenting that you have done so. (Of course, many of the things you do to maintain your marketability will also increase your value to your current employer, thus increasing your job security and prospects for career advancement.) Keeping up-to-date in your field means continuously learning about new advances in your area of science of engineering. Define the term "field" broadly, as some narrow specialties will fall out of favor. Narrowly based practitioners who do not maintain their skills in other areas of technology will face major difficulties in the job market.

Mastering relevant technologies also means understanding the impact of new analysis techniques on your specialty and mastering new communications and office technologies. For example, today's job candidate who has not mastered a computer, who cannot access the Internet to network with peers and post for job openings, and who cannot suggest appropriate occasions to use videoconferencing will face major handicaps in both getting and keeping a job. Knowing the business conditions in the industries in which you are job hunting and major management issues is also very helpful as you progress to employment interviews and discussions with professional recruiters. An established network of professional contacts (chapter 5) to rely on when you enter the job market is also important. Waiting until you enter the job market to try to build a professional network will place you under a major handicap.

KEEPING CURRENT IN RELEVANT TECHNOLOGY

Continuing education is key to keeping your professional qualifications up to date. Keeping abreast of technology in your field will help you develop new products and services for your employer to market. Staying up-to-date will enable you to know the advantages that your new development provides over current products, processes, or services. This will help your employer market your new development. You will be able to recognize when your new discoveries are inventions that your employer should patent by staying current in your field.

You should understand your employer's technology in all areas relevant to your current responsibilities. Should you wish a transfer to another area of technology, master the relevant information before asking for a transfer or before an opening becomes available. Demonstrating your knowledge can provide you with a competitive advantage over coworkers who mat be competing for the new position. It is also important to master your customers' technology. This knowledge will help you develop more effective solutions to their problems and even anticipate their concerns. Having a solution to a new problem "sitting on the shelf" and ready to go can really impress your customers, your employer's salespeople, and your supervisor or fellow team members.

Continuing education also means acquiring knowledge in new fields. Often, these are areas of technology where growth is occurring much faster than in your own specialty. This growth includes knowledge, commercial opportunities, and employment opportunities. Often your entrance into a new field can be based on your current expertise. For example, chemical engineers who learn about fermentation technology have an in to biotechnology when companies are seeking to scale up laboratory processes to pilot plants and commercial units. By combining a study of fermentation technology with some reading in biochemistry and biotechnology, the alert chemical engineer now employed in the petrochemical industry could switch fields, as well as employers. The engineer who works in the food industry where much chemical technology is based on fermentation has a head start.

Continuing education also means mastering new skills that support your core competencies. For example, scientists and engineers have had little choice but to master personal computer technology over the past 15 years. The improved productivity that microcomputers can provide may have begun as an option, but is now a necessity as everyone strives to be more productive. Personal computer graphics capabilities make it much easier to tailor technical presentations to your audience. Word processors make it possible to take a confidential research and modify it for sales representatives or a technical publication. I would never have the time for both chemistry and writing careers if word processors were not available to make the writing process so much easier.

Word processors and e-mail make it possible to maintain a wide professional correspondence while investing much less time than would have been required two decades ago. Thus, maintaining your professional network can be easier than before even if your professional and personal schedules are busier than ever.

Mastering computer software technology can improve your productivity while expanding your capabilities. Software can enable you to do molecular modeling and better understand and predict stereochemistry and kinetics of chemical reactions. It can also enable you to simulate processes as different as chemical plants, petroleum recovery from subterranean formations, and model biochemical processes in living organisms. Of course, most of us do not have to know how to write the programs. What we need to know is how to use them and how to pose useful questions and provide reasonable input.

Reading widely, despite a busy schedule, can enable you to take advantage of new technology and apply it to your work. For example, atomic force microscopy, which was developed little more than 10 years ago, is becoming a widely used analytical technique to study surface chemistry, interactions at surfaces, and surface characteristics of materials. It may be too late to become a pioneer in applying atomic force microscopy to understand and solve problems in your field; however, other technology breakthroughs that will provide new problem solving capabilities in many fields are happening now.

Oral presentation skills, writing skills, foreign language skills, interpersonal skills, time-management skills, teamwork skills, people-management skills, and career-management skills all support your basic technical competencies. These are reviewed in previous chapters. This is just a partial list; your own list may be longer. Many of these skills are not taught in colleges. Independent reading and short courses are usually the way to improve these skills. Sometimes organizations can help. Programs such as Toastmasters International can help you improve your oral communications skills. Writers clubs seldom focus on technical writing. However, taking a short course, diligently practicing its precepts, and using the grammar checker on your work processor can all help to improve your writing skills. The key is to practice what you learn from your short course or your reading and not fall back into your old habits. This is particularly difficult when it comes to people-management, interpersonal, and teamwork skills. In large part, our basic personalities determine how most of us practice these skills. It usually takes a life-changing event to change our personalities. However, one can still learn to improve these skills and practice them more effectively. It is often a hard thing to do—harder than mastering scientific and engineering skills.

DOCUMENTING YOUR ACCOMPLISHMENTS

Documenting your accomplishments is critical to establishing a reputation as a productive, results-oriented individual and team member. This is key to improving your job security with your current employer and improving your employment prospects should you enter the job market.

While your immediate team may know of your accomplishments through personal discussions and your oral presentations at team meetings, coworkers in other departments also need to know your accomplishments. They should think of you when forming a team to solve a problem or as a person from whom to seek advice. This is job security in the modern era of horizontal corporations, corporate restructuring, and frequent changes of projects. You may think your supervisor is quite familiar with your work. However, when the time comes for supervisors and team leaders to assess your performance, you want them to remember your accomplishments from several months ago. Documenting your work in corporate reports can help them do this and lead to a better raise or a choice assignment.

Other forms of documentation can both help provide these benefits to you and protect the rights of your employer to commercialize the technology you have developed. One way to do this is for your employer to file for United States and foreign patents on your inventions. The first step in doing this is often entirely up to the inventors' initiative. When you think you have invented something of value to your employer, prepare an invention disclosure. Having previously written good internal reports can help you do this. These reports will also help you and the patent attorney prepare the patent application. Some companies have patent agents or patent attorneys constantly review internal reports looking for inventions that should be patented. However, as a potential inventor, you should not rely on someone else to suggest that you write an invention disclosure. You are probably better equipped with the background knowledge needed to decide whether one of your discoveries is indeed an invention.

The decision whether to file a patent application is often an economic, as well as a technical, issue. Some patents may be difficult to enforce or easy to design around. In these cases, your employer may wish to keep your discovery a trade secret. In other cases, your employer may decide the economic benefits your invention provides may be insufficient to justify filing a patent application. Alternatively, they may decide that a patent on your discovery may be very difficult to obtain.

In these last two situations, you may want to publicly disclose your discovery. This will prevent someone else from independently making the same or a very similar discovery, patenting it, and preventing your employer from commercializing your discovery. This often is an important justification in persuading an employer to allow you to present technical papers at scientific or trade association meetings or publish papers in scientific and trade journals. This public disclosure of your information can have very beneficial effects for you personally. Coworkers become aware of your contributions. Favorable comments from peers, competitors, and customers to your coworkers are particularly valuable. Finally, the awareness of your peers who work for other companies, competitors, and customers, can help you develop an industry-wide reputation. This can be very useful should you decide to enter the job market.

◈ Employment Status Does Not Determine Inventorship

Anyone who makes an interesting discovery can be an inventor. This includes technicians who observe an interesting phenomenon in the laboratory or independently modify test procedures someone has given them and obtained a useful and unexpected result. The sales representative who listens to a customer's problem and comes up with a novel solution may also be an inventor. So is the plant engineer or operator who develops a novel solution to a plant problem. Do not let your corporate status inhibit you from submitting an invention disclosure.

However, do not file invention disclosure without careful thought. When you have serious doubts about whether to do so, consult a mentor or trusted coworker for their advice. You can also discuss the situation with your company's attorney. It is the patent attorney's job to determine whether a discovery is an invention when this is unclear to the inventor. This is often a legal question that bench chemists, engineers, and technicians may be unable to determine.

MAINTAIN A HIGH PROFILE OUTSIDE YOUR COMPANY

Developing and maintaining a reputation outside your company can be very valuable to your career. Coworkers who hear people from other companies make favorable comments about you are often deeply impressed. This will also help your career with your current employer. A good reputation outside your company will provide you with valuable network contacts should you make a job or career change. Some of these will provide job leads or new employment opportunities. Outside peers impressed with your reputation often can provide you with interesting professional opportunities and challenges. These include serving on professional society committees, chairing conference sessions, being asked to contribute a paper to a journal or author a chapter for a book, and so on.

So how can you begin to develop a solid reputation outside your company? Begin by networking aggressively at scientific and trade association national, regional, and local meetings. Also network at short courses. (Networking is discussed at length in chapter 5). Ask good questions of speakers during technical conferences. However, avoid seeming antagonistic or excessively critical. It will not happen after only one or two meetings, but asking penetrating, but civil, questions can be a very effective and rapid way to establish your name as a competent expert. Participating in scientific society and trade association activities and governance can add valuable contacts to your network. Programming activities in particular are an excellent way to meet the leaders in your field. My own experiences illustrate the value of this. In late 1989, my employer decided to develop and market products to remove ink from pulped wastepaper. The main technical organization that organized meetings and publisher papers in this field is the Technical Association of the Pulp and Paper Industry. I immediately became active in this organization and in 1991 organized the pulp de-inking part of the paper recycling program for an annual national meeting. Then, in 1993 and 1994, I organized the entire paper recycling program for this annual national meeting. As a result of contacting many experts in several fields to invite them to present papers, I got to know the leaders in paper recycling and they got to know me. As a rookie in the field, at first I did not have much technical insight to offer, but hard work on programming committees and the clearly evident results enabled me to rapidly develop respect. Soon I was a welcome member of technical discussions during conferences. Technology leaders took my calls and helped me to develop a knowledge base. Of course, reading helped, but these personal interactions greatly aided me in mastering the field. This has been beneficial as a result of learning experiences and opportunities that some of these leaders have presented to me. It has been beneficial to my employer when sales representatives call on prospective customers. Since some of their technical people know me personally, this gives my employer some instant credibility.

Of course, if you and your employer are to benefit from this sort of activity, you must do a good job. My own example above shows this. If you volunteer or are assigned professional society duties, make sure you fulfill your responsibilities competently and on time.

Accompanying sales representatives on their sales calls has many benefits. So can assisting sales and marketing personnel in staffing trade show exhibit booths. You will get to meet customers who will become familiar with your skills and capabilities. Solving problems for customers can turn them into valuable members of your network eager to help you when you need advice or other assistance. Of course, solving problems for the customer benefits your employer, as well, and can lead to increased customer loyalty and possible new sales.

JOB HUNTING

Maintaining your marketability increases productivity in developing leads in the job-hunting activities of distributing your resume and cover letter, networking, and interviewing. These subjects are discussed in the following chapters.

Additional Reading

Biracree, T. and Biracree, N. *Over Fifty: The Resource Book for the Better Half of your Life,* Harper Perennial: New York, 1991.

Bird, C. *Second Careers: New Ways to Work After Fifty.* Little, Brown & Co.: Boston, MA, 1992.

Intellectual Property

Elias S. and Goldoftas, L. *Patent, Copyright & Trademark: A Desk Reference to Intellectual Property Law* Nolo: Berkeley, CA, 1997.
Levy, R. C. *The Inventor's Desktop Companion: The Guide to Successfully Marketing and Protecting Your Ideas.* Visible Ink: Detroit, MI, 1995.

PART III

JOB
HUNTING

12

THE EVOLVING TECHNIQUES
OF JOB HUNTING

Just as the employment world has changed greatly for scientists, engineers, and technicians, so has the world of job hunting. These changes are continuing and no doubt there will be new developments. Among the developments of the last several years that have affected job hunting for scientists and engineers in a major way are:

- the widespread availability of personal computers, word-processing programs, and letter-quality printers. This has resulted in greatly increased use of targeted resumes and tailored cover letters (see chapters 13 and 14). Because of the widespread availability of this technology, employers now have a near-zero tolerance of typographical errors and poor grammar.
- computer-scannable resumes (see chapter 13)
- use of the Internet both by job hunters and employers wanting to fill job openings
- professional societies' increased role in assisting the job hunting efforts of their members by providing job banks, computer bulletin boards, and other services. Many professional societies are now using their home pages on the Internet to inform their members of these services.

Trends in their early stages of development that may have a major impact on job hunting include:

- use of e-mail as a means of communication between job hunters and employers
- employment interviews using video conferencing technology

One interesting thing about these developments is that, at least to date, they have not replaced traditional methods of job hunting. Instead, they have supplemented them. Midcareer scientists and engineers who have not job hunted in the last 10 to 15 years will face a major shock if they have not updated their knowledge of job-hunting technology and suddenly find themselves in the job market. Of course, this job-hunting situation has become much more likely in today's employment environment.

A general tendency of these developments is to decrease the time you have to provide an employer with information when they ask for it. Fifteen years ago, when an employer called and asked you for a copy of your resume, you put it in the mail and they would receive it in three to four days. Now the fax machine and e-mail have changed expectations. If employers ask for your resume or additional information, frequently they expect to receive it the same day. You have less time to collect requested information such as names and phone numbers of references or details of a particular assignment you performed on your last job. You need all this information prepared and at your fingertips.

INFORMATION MANAGEMENT DURING JOB HUNTING

The highly competitive job market means that your job hunting must be efficient and organized. Job hunting generates a lot of information. This can be particularly true for midcareer professionals whose job hunting spans several months. Information that seems impossible to forget now may soon enough fade. Often you need to access this information quickly, sometimes during a phone call to a prospective employer.

Using Your Personal Computer to Manage Information

You can use your personal computer to organize and store your job-hunting information. This information can include different versions of your resume and cover letter. Storing examples of your accomplishments and abilities is helpful. You can "mix and match" examples and anecdotes concerning your skills and accomplishments to target your resume and tailor your cover letter to specific job openings. For all this you will need a word processor and a letter-quality printer.

If your word processor does not have a spell or grammar check, you should consider purchasing separate programs to perform these functions. You can often miss mistakes, particularly when you are in a hurry to print a document and fax it to a prospective employer. However, do not rely solely on your spell and grammar check. You may have used an inappropriate word that, because it is in your spell check's dictionary, will not be identified as being misspelled. (The most famous example of this is using "there" when you mean "their.") Also, your review may enable you to improve awkward wording or poor style not identified by your grammar check.

If you do mass mailings of a resume and cover letter to multiple employers, a mail merge program can save you time in printing letter headings and addressing envelopes. You will be contacting prospective employers by letter, phone calls, and perhaps by fax and e-mail. You will be sending them different versions of your resume and cover letter tailored to particular job openings. You will also be calling members of your network to see if they have information on job openings and names of other individuals who might be able to help you in your job hunt. All these contacts generate a great deal of information. You can record this information for later retrieval using a database management program, spreadsheet software, or contact management software. (Sales representatives use contact management software to manage information on their calls on customers, topics discussed, information requested, and samples sent to the customer.) You can record information sent to prospective employers regarding when you sent it, the names and phone numbers of people you contacted, summaries of conversations, and dates when they occurred. You can retrieve this information to determine if you should write or telephone a prospective employer to determine the status of your application or just refresh your memory. Summaries of interviews, your answers and comments on tough interview questions, and impressions of interviewers can all be useful information to record and retrieve later.

A calendar and scheduling program can provide useful reminders of when to call prospective employers or members of your network. Recording your interview dates and travel arrangements can help you avoid scheduling conflicts and hectic rushes to the airport. I personally prefer a paper organizer with "month-at-a-glance" calendars, as well as a page for each date to manage my schedule (but maybe I'm just old fashioned!).

THE INTERNET

The Internet is adding new dimensions to job hunting. Many companies advertise their available professional positions on their home pages. Many professional societies have set up job banks for their members. On their home pages, many societies carry relevant employment advertisements from their own and other magazines. There are many job banks on line, some of which are organized by private groups and may or may not have fees. Others are sponsored by state agencies. Some of these carry classified employment advertisements from major newspapers. Some newspapers have their own on-line editions that include employment advertisements.

Internet developments continue to occur rapidly. Rather than produce a summary here that will become rapidly dated, the reader is recommended to use a browser to locate employment sites on the Internet.

RECRUITERS

Unlike employment agencies, recruiters collect their fees from employers, not job hunters. First we will talk about how recruiters work, then discuss about how you can take advantage of their contacts and abilities. Some technical professionals refuse to work with recruiters. This is short sighted as it limits your job search. However, it is wise not to work with a firm that claims to be a recruiter, but charges you fees for revising your resume or for placing you in a position. If you are desperate for a job and these firms are your last resort, check them out with your local Better Business Bureau first.

How Recruiters Work

There are two types of recruitment firms. Retained firms are paid regular fees by employers. They usually work to fill positions paying more than about $85,000 annually. In contrast, contingency firms are paid only when a candidate they provide accepts the job. They typically work on filling positions paying between about $30,000 and $85,000 annually. Both types of firms often specialize in certain functions or industries. If you are actively job hunting, contact firms that specialize in filling science and engineering openings.

Retained firms never disclose names of candidates to third parties. However, contingency firms may propose your name to several different companies. If your skills and experience match those of a job opening, they will send your resume to that employer. Of course, if you lack the necessary credentials, your resume does not go anywhere.

Recruiters often hold client (the employer is their client, not you) names confidential. So you may not know to whom your resume is being sent. If you want to keep your job search completely confidential, establish certain ground rules with the recruiter such as always contacting you before proposing you for an opening and never mass-mailing your resume to employers. When they call to get your approval to propose your name for an opening, but want to keep the client's name confidential, you can decide whether or not to contact the anonymous employer based on the attractiveness of the position and the strength of your privacy concerns.

What to Do When a Recruiter Calls

Never ignore a recruiter's phone call. If you are job hunting when recruiters call, they can of course be very helpful in locating opportunities. However, even if you are not, listen to what they have to say. After they describe a job opening, suggest possible candidates to them and provide phone numbers if you have them. Of course, I never recommend coworkers; that would be unethical. However, I keep a list of job-hunting colleagues from other firms. Also, I sometimes suggest good candidates who work for my employer's competitors. Recruiters usually remember you if you are polite and may call again when they have a job opening more suited to your skills and interests. They are much more likely to do so if you suggest a candidate. Recruiters can be powerful aids to a later job search when you are actively in the job market. They can be very helpful if you are employed and want to job hunt on a confidential basis.

OUTPLACEMENT FIRMS

Outplacement firms have become a common feature of the job market scene in the last decade. Many downsizing firms have hired them to assist former employees in writing resumes and polishing long unused job-hunting skills. Career counselors who work for these firms sometimes can discuss career options with the job hunter. Some facilities provide telephone, facsimile, word processing, and phone facilities to job hunters.

Of course, if your employer is paying, you do not have to worry about paying fees. Alternatively, you can contact an outplacement firm on your own and pay their fees. No matter who is paying, question the people you deal with to determine their credentials and experience. Even if they seem competent, do not rely on them as your sole source of information about how to write your resume and conduct your job hunt. Multiple sources of input lead to a more effective job hunt.

Job-Hunting Clubs

With the occurrence of widespread corporate downsizing, losing your job has lost much of its stigma. Professionals who used to job hunt alone can now benefit from the many job-hunting groups that have sprung up around the United States. These include national organizations such as "Forty Plus" for midcareer professionals. There are also groups sponsored by state agencies, churches, and civic groups. Others are informal groups that meet in churches or private homes.

As noted earlier, professional societies are providing more job-hunting services to their members. More organizations (including the American Chemical Society) are including job banks and resume posting services as part of their websites. Other services include job-hunting groups sponsored by local sections of professional societies such as the Chemical/Engineering Opportunities Office jointly sponsored by the Houston area local sections of the American Chemical Society and the American Institute of Chemical Engineers.

Some services that members of job-hunting groups provide each other include resume critiques, mock interviews, and evaluation of each other's interviewing techniques, and networking to learn of job openings. Some groups recruit expert speakers to discuss job-hunting techniques and employment opportunities in various industries. Emotional support is an important aspect of many of these groups.

COMBINING TECHNIQUES

As noted in the beginning of this chapter, new techniques of job hunting are not replacing traditional methods; they are supplementing them. There are many techniques to make you aware of job openings:

- employment advertisements in newspapers, professional society magazines, and trade magazines
- through corporate and independent recruiters
- through job banks on the Internet and those offered by on-line services. Some companies advertise job openings on their Internet home pages.

- networking
- mass mailing your resume and cover letter
- on-campus interviewing (mostly for students)
- interviewing at employment clearing houses and job fairs
- placing a "position-wanted" advertisement in appropriate trade magazines. You can usually do this anonymously. Several professional organizations that publish their own magazines provide this service free for unemployed members. These techniques seldom produce many employment interviews, but can be a useful adjunct to your job search.

By itself, no technique has a terribly high success average. To shorten your job hunt and find the most suitable job for you, you should use as many of these techniques as possible. However, before you begin, you will need at least one resume.

Additional Reading

Fegler, H. *The Complete Job Search Handbook*. Henry Holt: New York, 1988.

Krannich, R. L. and Krannich, C. R. *Change Your Job, Change Your Life: High Impact Strategies for Finding Great Jobs into the 21st Century*. Impact: San Luis Obispo, CA, 1996.

Messmer Jr., M. *Job Hunting for Dummies*. IDG Books Worldwide: Chicago, IL, 1995.

Weinberg, J. *How to Win the Job You Really Want*. Henry Holt: New York, 1989.

For Midcareer Job Hunters

Biracree, T. and Biracree, N. *Over Fifty: The Resource Book for the Better Half of Your Life*. Harper Perennial: New York, 1991.

Bird, C. *Second Careers: New Ways to Work After Fifty*. Little, Brown & Co.: Boston, MA, 1992.

Morin, W. J. and Cabrera, J. C. *Parting Company: How to Survive the Loss of a Job and Find Another Successfully*. Harcourt Brace Jovanovich: New York, 1991.

Porot, D. and Bolles, R. *The PIE Method for Career Success: A Unique Way to Find Your Ideal Job*. JIST Works, Inc.: Indianapolis, IN, 1995.

For Academic Jobs

Heiberger, M. M. and Miller Vick, J. *The Academic Job Search Handbook*. University of Pennsylvania Press: Philadelphia, PA, 1996.

For New Graduates Especially

Carmichael, M., Poole, L. L., and Hoffman, M. *Real Life Guide to Starting Your Career: How to Get the Right Job Right Now*. Pipeline: Carrboro, NC, 1998.

For Families of Job Hunters

St. Pierre, S. *Everything You Need to Know When a Parent is Out of Work*. Rosen: New York, 1997.

13

WRITING YOUR RESUMES

No, the title of this chapter is not a misprint. Except for some students, having only one resume will not suffice in today's highly competitive job market. You need to tailor your resume to pursue different types of job openings or the same type of job in different industries. Targeted (also called focused) resumes will be a major subject of this chapter. While much of this material is aimed at experienced technical professionals, graduating students at all degree levels will find valuable information throughout this chapter. Information of special interest to students beginning their professional careers is presented at appropriate points in the text and identified.

As noted in chapter 12, resume standards have never been higher. There are several reasons for this. The first is the highly competitive job market. The second is the widespread availability of computers, wordprocessing software and spelling and grammar checkers. A single misstep such as a typographical error can eliminate you from the competition. A third is that good written communications skills are now part of most chemists', engineers', and technicians' jobs (see chapter 3). Poorly written resumes and cover letters signal employers that the offending applicant may be poorly qualified. An alternative to writing your own resumes is to use a resume writing service. (See appendix 1).

Interviewing job candidates requires a significant investment in staff time and money. To reduce this investment, employers use resumes as a screening tool to exclude less-qualified candidates from further consideration. Job candidates must use their resume and cover letter to convince employers that they have the ability to produce results valuable to the employer. If they succeed, an employer will invite the candidate for an on-site interview. The importance of the resume has recently led to a booming business for resume writing services. However, job candidates can prepare their own effective resumes by following seven steps.

"The resume should read like a continued value-added odyssey which adds up to 'The firm is better off because of the projects I performed, and I can prove it in no uncertain terms.'"

Management consultant Tom Peters
in *Liberation Management*

STEP 1—PLANNING

Planning can ensure that your resume presents a complete picture of your interests, abilities, and accomplishments. Preparing a resume is in part a process of self-analysis. This self-analysis will also help you prepare for employment interviews. While a resume service counselor can guide the applicant through this analysis to a certain degree, the process is often a lengthy one. The more time a counselor spends with an applicant, the higher the fee. The applicant may find it personally more rewarding and educational to take this voyage of self-discovery alone.

Spend at least two planning sessions assembling a complete list of your accomplishments, abilities, and interests. For students, this should include college classes, relevant extracurricular activities, scholarships, awards, part-time and summer jobs, and community and volunteer work. More-experienced professionals should limit descriptions of their education to colleges attended, degrees awarded, and theses written. They should focus on their employment experience. Accomplishments, rather than duties and responsibilities, should be emphasized. Off-the-job professional accomplishments and activities should be described only briefly. Younger professionals with less than five years experience will need to strike a balance between their on-the-job accomplishments and their education. Describe situations in which you:

- demonstrated problem-solving skills
- showed that you were a team player
- exercised leadership skills
- increased revenues
- increased efficiency
- wrote outstanding reports

Also, try to remember when you demonstrated personality traits of interest to a prospective employer such as: creativity, perseverance, enthusiasm, loyalty, being a self-starter, and diplomacy.

◆ The Curriculum Vitae

The guidelines for resume writing and the curriculum vitae differ considerably. The curriculum vitae is used only in applying for academic positions and research grants. Unlike the resume which should be concise and only emphasize major aspects of one's career and accomplishments, the curriculum vitae is a long document that includes duties and responsibilities, as well as accomplishments, detailed descriptions of past, current, and proposed research, and lists of publications and references. It uses a chronological structure. While the resume leads to the next step of the industry hiring process (the on-site interview), the curriculum vitae does more than this. It is a major factor, usually *the* major factor, in the hiring decision for an academic opening.

Graduating students should consult extensively with their professors in preparing their curriculum vitae.

In assembling this list of activities and accomplishments, note the features of each you found most and least rewarding and enjoyable. Students, you should consult with your parents, teachers that know you well, and trusted friends to get their impressions of your strengths and weaknesses, skills and achievements. More-experienced professionals can consult with trusted peers and mentors. Also list and prioritize your personal values.

In reviewing this information, you may begin to see a pattern emerging—a correlation between the accomplishments you found most rewarding, people's perceptions of your abilities, and your personal values. These can help you focus on the professional areas for which you are best suited and will find most rewarding. Career areas for chemists, engineers, and technicians include:

- plant engineer or chemist, technician in a plant quality assurance laboratory, plant operator
- industrial basic research
- applied research and development
- technical service
- sales and marketing
- technical or general management
- faculty position at a research university
- faculty position at a four-year college

Consider carefully which options best fit your skills, abilities, and interests. For students, discussions with experienced professionals, recent graduates, and faculty members can help define which of these options is best. Experienced professionals should discuss their options with trusted peers and mentors. They should consider which aspects of their jobs they like the best. This may lead a basic researcher to conclude that she enjoys interacting with customers *and* working in the lab. She could combine both in an applied research or technical service position.

This self-assessment can be difficult in the emotional turmoil of losing your job. An annual career "check-up" (chapter 18) can help you make these assessments in a nonstressful way. By performing an annual career check-up, you can use your self-assessment to make career changes without necessarily having to enter the job market.

Writing your resume should not be done in a hurry. Experienced professionals can update their resumes as part of their annual career check-up (see chapter 18). It is helpful for students to begin planning their resume a year before graduation. Their self-assessment may lead them to identify a previously unconsidered area such as management or marketing as serious career options. If this is the case, it is often advisable to take courses relevant to these options in your final academic year.

Your planning is complete when you have selected one or two broad career areas, identified your personal abilities and accomplishments, and tabulated a complete employment and education history.

STEP 2—ORGANIZING YOUR INFORMATION

When you have completed your planning, you will have a large amount of information. A personal computer can be an invaluable aid in organizing this information and later in writing your resume. Begin by writing your job-hunting objective. If you ca not decide between two different areas such as academic or industrial basic research, write separate resumes for each. All the information in your resume and cover letter should show that you have the interests, skills, accomplishments, and personality traits to function well in the type of position that is the target of your job hunt. Since you may have more than one target, you may need more than one resume. This is true for students, as well as experienced chemists, engineers, and technicians.

The best organization of this information in your resume may differ for a graduating student and an experienced professional. The most common resume

---

◆ Resume Formats for Graduating Students

Graduating students usually use the chronological format for their resumes. It serves to identify their educational progression. If the student completed his education in a very short time or attending a particularly prestigious university, the chronological organization helps him emphasize this. However, in many circumstances, a functional resume will be more effective and help graduating students stand out from a crowd of applicants. A functional organization lets them emphasize an in-depth exposure to a field or technology in both graduate and undergraduate school. If a summer job or a job during a hiatus between undergraduate and graduate school is particularly relevant to the type of job they want, a functional organization that allows the student to place these job skills in prominent location is best.

format is chronological. Both your education and work experience is presented in reverse chronological order. Thus the emphasis is on your most recent accomplishments. An alternative is to use a functional organization in which you organize your experience by skills and accomplishments rather than chronologically.

The chronological format is usually preferred for those who have had a long career in a single technical field and whose employment history is a record of increasing accomplishment and a series of promotions. These individuals have a single set of skills that served them well in a series of similar jobs or assignments. The chronological format emphasizes a continuous record of employment.

The functional organization allows the candidate to put the most relevant information first in a prominent location. This resume format is well suited to being modified and tailored to a specific job opening. This organization is often preferred for an experienced candidate's resume. It allows the candidate to present all the information on each skill in one place. This is helpful when the candidate has a broad range of skills and the skills most relevant to the job opening were practiced most in an earlier assignment.

The functional resume also is preferred for candidates making a career change and wanting to emphasize skills transferable to their desired new careers. It is also preferred for those who cannot document continued employment and accomplishments. This includes new graduates, those with interrupted careers, and those with employment in nontechnical areas.

Examples of resume formats are presented in Appendix 2. Resume-writing computer software often provides templates of various resume formats. This makes it easier to draft your resume and compare the effectiveness of different formats. You can do this as part of Step 5 discussed below. Do not blindly follow the advice of the references listed at the end of this chapter or that of a resume-writing service. Choose your resume format carefully to be sure it is the most effective one for you.

While experienced professionals should begin with their work experience, graduating students should start with their academic credentials and achievements. Note your college degrees and briefly describe your major field. Mention special courses and seminars that were both particularly valuable and could distinguish you from other applicants. An English or Journalism minor, foreign language fluency, or programming courses could provide you with special skills most applicants do not have. Graduating college students should not include high school information except to note if they graduated with honors or won scholarships. Other details of your high school are already ancient history. However, students obtaining advanced degrees can find much in their undergraduate education that can be helpful in their job hunt. Activities in student chapters of professional societies can also be helpful.

If you have special computer skills, note them. However, mere mastery of word processing, spreadsheet, database, and other software is no longer so unusual as to be of special interest. To cover this situation you can briefly mention that you are "IBM PC literate." Students should include special projects and extracurricular activities that demonstrate relevant skills, accomplishments, and personality traits. Experienced professionals should mention extracurricular activities if they have been out of school for less than three years. Students should focus on activities that indicate leadership and teamwork skills, identify you as an achiever, and illustrate your values. Use titles—president, chairperson, secretary, and so on—that you held while involved in these activities. Detail the number of people who worked for or with you on these projects or activities. Do not become carried away listing too many extracurricular activities; you do not want to appear to have majored in fun and games.

Place any mention of hobbies and special interests at the end of the resume. However, work on writing them while describing your special skills and activities since you are in the proper mindset. The only reason to mention hobbies and special interests is so you can briefly describe valuable work skills learned from these activities. One example is learning teamwork and patience in accomplishing long term goals as a member of a model railroad club.

When writing about work experience and achievements, students should not worry if their student and summer jobs do not sound sophisticated. They demonstrate your commitment to finance a part of your education. They provide information about your values, motivation, maturity, and work skills such as sense of responsibility, teamwork, and leadership.

In describing employment history, both students and experienced professionals should start with their most recent job and provide the name and location of each employer, job title, activities, responsibilities, and accomplishments. Students should try to do this in six lines or less. Experienced professionals may wish to devote up to about eight lines on their most recent job. If experienced job hunters use a skills-oriented resume (see appendix 2), employment history is briefly listed with employment dates, employers, and job titles detailed. Describe your accomplishments and skills in a separate section. Skills-oriented resumes

◆ Avoid Overweight Resumes

Job hunters often cram too much information into a resume because they want it to substitute for an interview. This strategy seldom works. Follow the LLOO Rule: keep your resume lean, logical, orderly, and objective. Have your resume reviewed by others to help you limit its length.

are usually targeted resumes. Your emphasis on these various accomplishments and skills is determined by the type of job for which you are applying.

STEPS 3 AND 4—EFFECTIVE WRITING AND FORMAT

You are now ready to take the written descriptions you prepared in Step 2 and prepare a draft of your resume. It helps to have a format in mind when you start to write. Even though an unusual format may be eye-catching, do not get too avant garde. The two most important format features are a logical organization and presentation of information and plenty of blank space. The blank space serves to both make the organization clear and reduce the intimidation large blocks of text have on readers.

Many employers first rapidly skim resumes to eliminate candidates unsuited for the available position. Your resume may have to survive as many as four levels of screening: an employment secretary, an employment staff assistant, an employment manager, and hiring authority. Since many organizations receive thousands of resumes in a year, everyone but the hiring authority may spend as little as 60 seconds scanning your resume. (If a firm is not currently hiring, your resume could be discarded or scanned into a computer [see below] without being read at all!) An effective and attractive format will increase the odds that your resume will survive the preliminary screenings and reach the hiring authority.

With the format and organization in mind and your information assembled, you are now ready to write. The resume and its cover letter should be examples of your best writing. This will be your first, and possibly only, opportunity to demonstrate your written communication skills to a prospective employer. Poor writing can overwhelm otherwise excellent credentials. Schedule a large block of time for writing. Even with all your advance preparation, you will probably need at least an afternoon for the first draft.

Before beginning, remember your goal is to define yourself as an intelligent, hard-working leader and team player, who is a good communicator, a fast starter, and someone willing to perform at the peak of their abilities. You must sell your skills and experience as they relate to the job you want. Doing all this is a tall order.

Be concise. Most experts recommend that new graduates and people with less than five years experience limit resume length to one page. Do not sacrifice an attractive appearance to cram information onto a page. An experienced applicant's resume should be no more than two pages long. Submit relevant supporting information such as a publication list on a separate page. Do not staple this to the resume because the person reviewing the resume may think your resume is more than one page.

Write a job objective at the top of the resume. Do not include a job title since these can differ greatly among companies. The job objective can indicate whether there will be both a short-term and a long-term fit between your goals and those of the prospective employer. It should be focused and detail the following:

- the type of work you want
- the size and type of firm you want to work for
- geographic restrictions

The job description should be brief, less than three lines long. Many employers use the job objective to screen resumes because they do not have time to read each resume in its entirety.

◆ What Does the Resume Reader Want to See?

Resume screeners may want to know if your qualifications meet those specified in a job opening advertisement. They also need to quickly decide if your qualifications could meet the requirements of any departments with positions available. If their answer to this question is favorable, they will forward your resume to the appropriate department. If your qualifications may be relevant to a possible future opening, they will keep your resume on file, perhaps scanning it into a computer database (see below). If they are uncertain about the relevance of your qualifications because your resume is poorly written, they will probably discard it. To avoid this terrible fate:

- Address the employer's needs and priorities, not your career aspirations.
- Provide benefits. Within the limit of nondisclosure agreements with previous employers, provide quantitative information. Numbers impress. Numbers enable readers to better judge the significance of an accomplishment. The statement, "As plant engineer, I headed the team that debottlenecked a distillation unit increasing plant capacity by 20%" will impress the reader far more than writing, "I headed the team that successfully debottlenecked necked our distillation unit."
- Do not tax the reader's patience with irrelevant information.
- Give resume screeners the information they want organized so they can forward your resume to the appropriate hiring authority quickly and not have to puzzle over what department can best benefit from your skills.
- Give hiring authorities the information they need to decide if there is a good chance you can fulfill the job requirements. After comparing your resume to those of other applicants, they can decide whether to invite you to an employment interview. Any information that is vague or confusing can send your resume to the bottom of their pile (if they do not just discard it).

Use clear phrases, rather than sentences. Start these phrases with action-oriented verbs such as:

analyze	approve	coordinate
create	delegate	demonstrate
develop	direct	discover
establish	influence	manage
plan	produce	write

These verbs connote leadership and productivity traits. Also use powerful adverbs to start phrases:

aggressively	effectively	energetically
equitably	objectively	quickly
reliably	successfully	tactfully

These adverbs indicate leadership, teamwork traits, and motivation. Be careful, it is easy to overdo the use of these adverbs.

If you feel you must use sentences, avoid overuse of the personal pronoun "I." Use the active voice instead. Using phrases, rather than complete sentences makes it easier to avoid using "I" and the passive voice too frequently. Do not mention irrelevant personal data such as height, weight, state of health, marital status, and so on. Military service is relevant. Include it as a separate category or in your previous employment section. Do not include a photograph with your resume.

Establish a clear organization and use your format to enhance it. Section headings should be concise and logically positioned. Your resume should be visually attractive. Use expanded or bold face type, underlining, bullets, and other highlighting techniques in moderation.

STEP 5—REVIEWING YOUR RESUME

At this point, you should have an attractive draft of your resume. Before you send your resume to anyone to review, ask the reader if the following questions can be answered:

- Is your product statement clear? What are your technical and/or managerial accomplishments? Are your current job level and responsibilities clearly described?
- What did you accomplish with those responsibilities?
- Do you clearly identify your previous employers, dates of employment, responsibilities, and accomplishments? (Trying to disguise lengthy periods of unemployment is seldom successful.) An employment period of three years or more suggests some productivity in your job assignments. Twenty years with the same company, particularly without clearly defined career advancement and increasing responsibilities, can suggest a marginal per-

◈ Listing References

Should you provide a list of references and phone numbers on your resume? This section will occupy some of your limited space. The alternative is a brief line at the end of your resume, "References provided upon request." If a hiring authority is interested in your qualifications, he is unlikely to discard your resume because you do not include names and phone numbers of references. By contacting you for this information, employers thus keep you informed of the status of your application. However, if you are difficult to contact, it is best to furnish telephone numbers of your references. A busy manager may be unlikely to try to call you more than once or twice.

After an employer asks for this information, warn your references to expect a telephone call (see chapter 15). By describing the position to them, your references will be better prepared to answer questions and describe your relevant qualifications.

former without much ambition. Periodic promotions and bonuses indicate high performance, as do interesting and significant-sounding job assignments, such as team leader of an important project.
- Where did you go to school?
- What degrees did you obtain?

After you review your resume, take it to family members, trusted friends, recent graduates, or professors who know you well. Students should consult a counselor in their college placement center. Students should find a recent chemistry or engineering graduate who has the type of job he desires and ask to see their resume, for comparison.

Experienced professionals may feel they know how to effectively write resumes. However, as someone who has written many resumes for others, I know that it is much more difficult to write my own. Even experienced professionals who are good writers can benefit from asking a trusted peer or mentor to review one or more of their resumes.

STEP 6—REVISING YOUR RESUME

Be prepared to be shocked and discouraged at this point. You asked for revisions and your resume could well be returned covered with red ink. Do not be close-minded. Consider each change your advisers recommend. However, do not follow their suggestions blindly. Discuss with them their recommendations if you have doubts; determine their reasoning and be sure it is valid. Revise your resume with the final format in mind. Try to boil it down to essentials when revising. Fight the tendency to add more information and significantly lengthen the resume. Also do not assume this draft will be your last. Ask again one or two people who previously had the most constructive comments to review your new draft. Then, make any additional changes that will improve the resume.

Analyze and quantify each of your cited accomplishments for business impact. Later, when you begin to receive employment interviews, listen carefully for frequently asked questions. When you identify one of these, check your resume to see if your answer is there. If not, revise your resume to provide this information and that it is presented in an appropriate location on the resume.

STEP 7—TYPING THE FINAL DRAFT

Use good-quality white bond paper. Light gray or very light tan was acceptable in the past. However, since many employers will scan your resume into a computer (see below), white is preferred. Some experts suggest that paper sized 7.5″ × 10.5″ is more noticeable and thus more effective than standard 8.5″ × 11″ paper. If you do vary your paper size, do not use a paper larger than standard size, since it cannot be stacked neatly. Larger paper may mean special adjustments need to be made to scan your resume.

Use a letter-quality printer to prepare the final draft. Be sure your resume is completely free of errors in spelling, typing, punctuation, and format. Proofread it carefully at least twice and have a friend do so as well. Do not mail photocopies. Have copies of your resume printed or print them yourself using a word processor.

You have done a lot of work, but you may not be done yet! While your cover letter can aid in targeting your resume to a specific job opening or a particular type of company, you should also consider revising your resume to tailor it for a specific industry or job opening. While a targeted resume is most valuable to an experienced candidate, new graduates can also find it of value. For instance, many chemists, chemical engineers, and technicians work in the petroleum, pharmaceutical, metallurgical, food, and other industries, as well as in chemical manufacturing. A candidate may want to prepare a targeted resume aimed at one or more of these industries. Experienced candidates can emphasize aspects of their experience and skills most relevant to a particular industry. Recent graduates can do the same for course work.

When preparing a targeted resume, do more than just add a section to your standard resume. You may wish to change your resume from a chronological format to a skills-oriented format. Identify your skills and experience that might be of particular relevance to the target industry and emphasize these in your resume and cover letters. Be particularly careful when defining your job objective so that you demonstrate that you will fill their needs.

COMPUTER-RETRIEVABLE RESUMES

Large, well-known firms may receive hundreds or even thousands of resumes annually from chemists, engineers, and technicians. All this paper is bulky to store. Organizing it in a systematic fashion is very time-consuming. Searching for candidates with particular qualifications is often tedious and inefficient. To cope with the sheer physical volume of all the paper, employers seldom hold paper resumes longer than one year. Chemical, petroleum, and pharmaceutical companies are among the leaders in setting up their own computer resume databases using optical scanners.

Many smaller companies receive relatively few resumes directly. As a result, such firms often have to contact technical recruiters to locate properly qualified candidates. This can be expensive for small firms with limited financial resources. However, these firms can access databases offered by independent resume data banks (see below).

Computer storage and retrieval of resumes solve these problems and benefit both employers and job hunters. Surveys indicate more than half of all companies with more than 500 employees are now using resume scanning and computer-storage technology. Many smaller firms are also adopting this technology. As a result, job hunters should design their resumes for optical scanning, storage in a computer database, and retrieval by keyword searching.

TAILORING YOUR RESUME FOR COMPUTER STORAGE AND RETRIEVAL

Reader-friendly does not automatically mean computer-friendly. Computers search by keywords or phrases. These should be common technical terms specific to your field of chemistry. Many job counselors advise against the use of technical terms. However, judicious use of technical terms is necessary if the computer is to find your resume in a search for a particular technical specialty. To aid computer searching, resumes can justifiably contain more technical terms than advisable in conventional resumes. However, remember that a term only needs to be used once for the computer to find it. Thus one should not make a resume bulge with obscure technical language. Remember, after the computer discovers your resume in a search, it still has to be read.

Chemists should avoid technical jargon and acronyms specific to their current employer. These will not be used as search keywords. Vague descriptions of job responsibilities are useless in retrieving your resume from a computer database. Since keywords are almost always nouns, the action-oriented verbs and adverbs discussed above are not helpful in computer retrieval. Preparing your resume for computer retrieval does not negate any of the rules discussed above, with one partial exception.

That exception is resume length. Since the computer search has already identified your resume as relevant to a job opening, your first reader will scan the resume more carefully than would otherwise be the case. This reader will be more tolerant of a longer resume. Use two pages if necessary to describe your skills and accomplishments. However, make sure your resume is tightly written following the advice given above.

COMPUTER-RETRIEVABLE RESUME MECHANICS

Format is critical for successful optical scanning of your resume into a computer database. Few firms will rekey your resume if it does not scan properly. Because optical scanners read from left to right, you should avoid double-column formats.

Unusual fonts, italics, underlining, and intricate graphics do not scan well. (These methods of getting a reader's attention are less necessary when your resume is picked in a computer search. The mere fact your resume was chosen means your qualifications are probably relevant to the job opening. Therefore, you already have reader interest.)

Use a standard font. The most common fonts are New Roman, Helvetica, and Courier. The optical character recognition software translates your resume into ASCII characters. Therefore, since all resumes retrieved will later be printed with the same font, there is little point in using an unusual font. It is also best to restrict yourself to one font for the entire text of the resume.

A type size of 12 points is preferred, but 10 points is acceptable. You may wish to use a 14-point type for major headings. Small type is difficult to scan. Similarly shaped letters such as "c" and "e" can look the same to a scanner if they are very small. To highlight section headings or specific accomplishments, use boldface type. Boldface type may scan into a computer as normal type. However, it will not interfere with storage of a clear image. Optical clarity of your resume must be excellent for successful scanning.

Creases across printed lines obscure clarity. Do not fold your resume when you mail it. Use a 9″ × 12″ envelope, rather than a standard Number 10 business envelope. (You may wish to stiffen the envelope with a piece of cardboard. However, this can increase postage.)

THE ROLE OF RESUME DATABASES

Placing your resume in one or more computer database should be only one part of a comprehensive search strategy. It supplements but does not replace other, more traditional methods of job hunting. This technology can actually enhance the effectiveness of mass mailing resumes and cover letters. Rather than dis-

◈ Independent Resume Data Banks

Some professional societies offer members resume listing in a computer database. The American Chemical Society is a leader in this field. The Online Professional Databank is a computerized service for members. Rather than supplying ACS with a conventional resume, members complete a detailed form on-line, which describes their qualifications, education, and prior employment history. Then, this data is entered into a computer. This nonconfidential service is free to members, national affiliates, and student affiliates, whreras for an annual fee, this service is available on a confidential basis.

Employers can search the database to find candidates with the qualifications needed. In addition, ACS has joined with Career Placement Registry to establish a computerized job matching service.

Many independent resume data banks also ask job hunters to complete a form rather than use their resume. This enables employers to more easily compare the qualifications of different candidates.

Some colleges are offering inclusion of their students' resumes in computer databases. Usually, there is no charge for this service.

Some services are available on-line. An example is the ACS Job Bank that became available on the World Wide Web early in 1996.

carding them, employers can scan them into a computer. They will probably be filed in this fashion longer than paper copies would be. Later, rather than inefficiently searching through a thick stack of resumes, a computer search will find qualified candidates quickly and easily. If the candidate's qualifications are appropriate and her resume well designed, the employer will find her resume when searching the computer database.

Most of the guidelines for computer-retrievable resumes apply to resumes used for job hunting on the Internet.

FINAL REMARKS

Writing a resume is a lengthy process requiring much. However, all this effort will be invaluable in your job hunt. When you get that great job, keep your resume, if possible on a computer disk. Your job experience and your old resume will make writing a new one much easier when you decide to look for your next job. Periodically updating your resume will let you take advantage of unexpected employment opportunities and accelerate your next job hunt.

A skills- or accomplishments-based resume may be the most useful during your job hunt. However, once on the job, it may be helpful to keep a chronological-format resume of your accomplishments. This can be useful in preparing for performance reviews and determining whether your career is gaining or losing momentum.

No matter how well written your resume, it is difficult for your personality to come through. Also, even if your resume content is impressive, it is difficult for a well-written resume to impress its reader with its style and writing excellence. These are two of the important roles of the resume cover letter.

ADDITIONAL RESOURCES

Many books and magazine articles have been written on resume writing and more continue to appear. What follows is a small selection of what is available.

General

ACS Career Services Booklet Tips on Résumé Preparation. American Chemical Society: Washington, DC.

Fein, R. 101 Quick Tips for a Dynamite Résumé. Impact: San Luis Obispo, CA, 1998.

Grappo, G. J. and Lewis, A. B. How to Write Better Résumés, 5th ed. Barron's Educational Series: Hauppauge, NY, 1998.

Jackson, T. and Jackson, E. The New Perfect Résumé. Main Street Books: New York, 1996.

Krannich, R. L. and Krannich, C. High Impact Résumés & Cover Letters: How to Communicate Your Qualifications to Employers, 7th ed. Impact: San Luis Obispo, CA, 1998.

Lain Kennedy, J. Résumés for Dummies. IDG Books: Chicago, IL, 1998.

Shuman, N. Revising Your Résumé. Wiley: New York, 1986.

Smith, M. H. The Résumé Writer's Handbook, 2nd ed. Harper Perennial, a Division of HarperCollins: New York, 1994.

Tepper, T. Power Résumés. Wiley: New York, 1998.

Yate, M. Résumés That Knock 'em Dead, 3rd ed. Bob Adams, Inc.: Holbrook, MA, 1997.

Resumes for Midcareer Job Hunters and Career Changers

Cochran, C. and Peerce, D. Heart & Soul Résumés: 7 Never-Before-Published Secrets to Capturing Heart & Soul in Your Résumé. Consulting Psychologists: Palo Alto, CA, 1998.

Marcus, J. J. The Résumé Doctor: How to Transform a Troublesome Work History into a Winning Résumé. HarperCollins: New York, 1996.

Rice, C. S. The $100,000 Résumé. McGraw-Hill: New York, NY 1998.

Rodmann, D., Bly, D. D., Owens, F., and Anderson, A-C. Career Transitions for Chemists. American Chemical Society: Washington, DC, 1995. See chapter 5.

Wilson, R. F. Better Résumés for Executives and Professionals. Barron's Educational Series: Hauppauge, NY, 1996.

Resume Writing for the Internet

Lain Kennedy, J. *Hook Up, Get Hired! The Internet Job Search Revolution.* Wiley: New York, 1995.

Lain Kennedy, J. and Morrow, T. J. *Electronic Job Search Revolution.* Wiley, New York, 1995.

Lain Kennedy, J. and Morrow, T. J. *Electronic Résumé Revolution.* Wiley: New York, 1995.

Weddle, P. D. *Electronic Résumés for the New Job Market.* Impact: San Luis Obispo, CA, 1994.

Weddle, P. D. *Internet Résumés: Take the Net to Your Next Job.* Impact: San Luis Obispo, CA, 1998.

Curriculum Vitae for Academic Positions

Anthony, R. and Roe, G. *101 Grade A Résumés for Teachers.* Barrons Educational Series: Hauppauge, NY, 1998.

Anthony, R. and Roe, G. *The Curriculum Vitae Handbook: How to Present and Promote Your Academic Career.* Rudi: San Francisco, CA, 1998.

Jackson, A. L. *How to Prepare Your Curriculum Vitae.* Vgm Career Horizons: Lincolnwood, IL, 1998.

Videos

"Developing the Right Picture: Résumé Preparation." American Chemical Society: Department of Career Services, Washington, DC.

"Formula For Success: Turning Job Leads into Gold." American Chemical Society: Department of Career Services, Washington, D. C.

Resume-Writing Software

WinWay Résumé Version 4.0 for Windows, Winway Corp., Sacramento, CA.

PFS Resume & Job Search Pro for Windows, The Learning Company, Cambridge, MA

The Perfect Résumé, Davidson Software Systems, Inc., Torrance, CA.

APPENDIX I

Resume-Writing Services

Hiring a resume service is the most expensive option to prepare a resume. Typical costs are from $40 to $500 and depend on the city and the amount of service you receive. Some resume services will merely take information you provide and return it in an attractive format. Others include career counseling and other job-search services.

Who should use a resume service? Students have straightforward needs and a relatively simple employment history. Counselors at a college placement office can often provide helpful advice on resume preparation. It is difficult to see how a resume service can provide value to students for the fees charged. Midcareer professionals may prefer to use a service to help them prepare resumes to deal with complicated employment histories and changing career interests. Resume services often prepare resumes that are more forceful in selling job candidates than candidates are able to write on their own. A skilled counselor can be more objective in writing the resume than can the job candidate. This can improve the selection and description of accomplishments in the resume.

Mid-career professionals may need someone with whom they can discuss their career plans on a confidential basis. If their professional network is limited, a resume-writing service could fill this gap. Counseling services offered by some firms can be particularly valuable if the job candidate is changing career interests. Another reason to use a service is if a chemist was not job hunting when she learned of an excellent employment opportunity. Using a resume-writing service may be the only way for her to respond promptly if she has not kept an updated resume filed in her computer.

Quality varies widely, so be careful in selecting a firm to prepare your resume. Services often advertise in local newspapers and telephone "Yellow Pages." Services, particularly those that specialize in helping scientists and engineers, advertise in professional society and trade journals. The experience of these services can improve the communication process and the resume they produce. Their specialized skills are usually reflected in the fees they charge.

Try to locate a resume-writing service in your own area so you can visit with them. However, your first contact should be by telephone to determine:

- What are the service's credentials? How long have they been in business? Do they have experience in preparing resumes for scientists, engineers, and technicians?
- Can you have an initial no-fee consultation with the person who will write your resume? What is this individual's background and experience?
- How will they go about the business of creating your resume?
- What is the fee structure? What is the method of payment?
- Will your resume be stored on computer so it can easily be updated or targeted to a specific job opening? What is the fee for these services?
- How will the service print your resume? What is the print and paper quality? How many copies of the resume will they provide and at what cost? What are the fees should you request additional copies later?
- Can you see samples of their work? These should be originals, not photocopies.
- What is the turnaround time? This is critical if you are trying to capitalize on a sudden, unexpected employment opportunity.
- Suppose the service introduces an error such as a misspelling? Will they correct it and provide fresh copies at no cost?
- Will the service also prepare cover letters?
- What other services such as interview counseling does the service provide?
- Can the service provide you access to computerized employer databases for a targeted resume mailing campaign?

After you get answers to these questions via a telephone conversation or no-fee personal interview, narrow down your choice of services to two or three.

Contact your state attorney's office or state consumer protection agency. These organizations can tell you if they have received complaints about particular services and the nature of those complaints. Another possible source of useful information is your local Better Business Bureau. These checks can be important in avoiding poor quality or unethical services.

A personal meeting is preferred for the complicated process of assembling and organizing information. Many firms provide a no-fee initial consultation. In this initial consultation, the services they provide and the fees they charge are explained. They also determine what services you will require. You can perhaps negotiate price. In particular, you can assess if you can work productively with the person assigned to you. (Most resume-writing services are single-person operations.)

It is also important to examine their previous work carefully. Do their resumes contain generic wording and vague phrases or more commanding and specific phrases? Choose a resume service that has a good track record, provides a high-quality product, meets your needs, and provides a compatible person for you to work with. Before beginning work, agree upon a format for your resume. When you begin, guard against faulty advice such as exaggerating accomplishments or using generic wording that reduces the impact of your accomplishments.

The completed resume should meet the LLOO Rule (lean, logical, orderly, and objective) and be visually pleasing. Be sure the format and printing, as well as the content, are to your liking before accepting the resume.

Alternatives

College placement center counselors advise students on how to improve their resumes. Midcareer professionals can consult trusted peers and mentors to receive similar advice. American Chemical Society career counselors will provide resume advice and review resumes free of charge at the ACS National Employment Clearing House held in conjunction with national meetings. The Regional Employment Clearing House is often held in conjunction with regional meetings. The services of ACS career counselors are often available at these meetings. You can also call the ACS to request the help of a career counselor. Other professional societies offer similar services to members.

APPENDIX 2

Examples of Resumes and Resume Formats
An Analytical Chemistry Student's Resume

As you read Florence's resume, note she did not exaggerate her accomplishments, but very clearly indicated she was busy on campus completing a senior research project, a minor in biology, working part-time for two years, and holding club offices. Her resume makes it clear that she is a smart chemist (making Dean's List and presenting her senior thesis research at two conferences) with good people skills (active in clubs and being elected to office). She worked two summers each at the diner and the savings and loan, which shows she was a good employee worth rehiring (although returning to previous employment may suggest some lack of sense of adventure).

Florence listed her work experience slightly out of chronological order in order to save space and limit her resume length to one page. She also used a tactic unusual for a student—emphasizing her accomplishments in a separate section and placing it in a prominent position in her resume.

Florence also used several terms that may stump a human resources person, but serve as good keywords for computer searchers: gas chromatography, mass spectrometry, MS, GC/MS, MALDI, FAB, alkoxylate, surfactants, and polymers. In discussing what were probably mundane duties in her chemical stockroom clerk job, she used the action verb "managed" when describing her record keeping duties rather than saying she "kept records"—a less impressive description.

Although Florence is a student and, therefore, might be expected to use a chronological resume, her natural lack of technical experience and the nature of her summer job experience led her to place her accomplishments and skills before her work experience.

FLORENCE FLASK
3333 Anywhere Lane
Bison, NY 20734
Tel.: 407-555-8764
e-mail: j_a_student@service.com

OBJECTIVE   Analytical chemistry position working with research and plant chemists in solving problems. Would eventually like the opportunity to work in both production plant and R&D environment.

ACCOMPLISHMENTS

Senior thesis presented as poster at the 219th ACS National Meeting of the American Chemical Society and at the Northwest New York ACS Meeting in Miniature

Dean's List eight semesters

SKILLS   Experienced in chromatographic, MS, and GC/MS methods of analysis. Senior Thesis: "Use of MALDI and

FAB Methods to Determine Molecular Weight and Molecular Weight Distribution of Alkoxylate Surfactants and Polymers."

Good at working with others. Led team that developed teaching lab safety inspection procedures.

WORK HISTORY

9/98–present	Chemistry Department stock room clerk, Bison State University. Maintained stock room in safe and orderly condition, transferred chemicals from large containers to smaller bottles suitable for laboratory use, managed records of chemicals signed out and returned to the stock room.
5/98–8/98 & 5/97–8/97	Waitress, Kozy Truck Stop Diner, New Jersey Turnpike, NJ. Standard waitress duties. Won Employee of the Month Award for July and August 1998 and July 1997.
9/97–5/98	Worked part-time as cashier and clerk in the Bison State Univ. Bookstore.
5/96–8/96 & 5/95–8/95	Teller, Hometown Savings & Loan, Hometown, NJ.

EDUCATION

B.S. Chemistry, Bison State University, expected 5/99. Minor in biology (21 credit hours). Took graduate-level courses in chromatographic analysis and mass spectrometry.

AFFILIATIONS

American Chemical Society Student Affiliate Chapter— President, Secretary Alpha Chi Sigma—Chapter Recorder, Vice President/Pledge Master

REFERENCES

Available upon request.

A Chemical Engineering Student's Resume

This resume is that of a chemical engineering student, Richard A. Chemeng, who participated in a co-op program that provided him two jobs with excellent industry experience. As a result, he has much more to talk about than did Florence in the "Experience" section of his resume. Wisely, he is trying to capitalize on this experience in his search for a job as a plant engineer.

Careful readers (who may not include the person screening the resume, but should include the hiring authority) will note that not only did Richard work for a semester for Washington State Chemical Company, but that they also asked him to work over the summer. While this may have delayed his graduation by a semester, it shows that he did a good job for the firm. Richard could emphasize this in his cover letter and note that the income enabled him to better finance his education.

Note also that Richard took some courses in biochemistry and biotechnology that could make him attractive to the burgeoning biotechnology industry. His co-op semester spent working in a fermentation plant (a brewing company) should also be helpful in obtaining a biotechnology job. Note also that he took a course in fermentation chemistry at another university while employed with the brewing company. This both helped him understand the underlying chemistry of his co-op job better and made him a more-attractive candidate to the biotechnology industry. He realizes this and includes biotechnology in his "Objective" statement. His experience could be very valuable to a biotechnology company seeking to scale up manufacture of a product and commercialize it.

Richard's 1996 summer job suggests that he is quite entrepreneurial, an advantage in the job market. Depending on the job opening, he might want to emphasize this in his cover letter. While he does not mention business in his resume, taking a couple of business courses might make him a strong candidate for a junior marketing or sales position that leads to a business management career.

Note that while Richard did not use a lot of action verbs in his resume, he has included many nouns excellent for keyword computer searching once his resume is stored on computer. In particular, there are a good amount of terms related to the biotechnology industry.

RICHARD A. CHEMENG

Anderson Hall, Room 227
University of Washington
Seattle, WA 99024
Tel.: 219-555-1111   e-mail: rachemeng@uwash.edu

OBJECTIVE

Plant engineer position in the chemical, pharmaceutical, or biotechnology industries. Plan to take business courses and eventually move into plant and business management.

EDUCATION

B.S. Chemical Engineering, February 2000, University of Washington GPA 3.58
Spent two semesters working in chemical plants as part of engineering co-op program. Standard engineering and chemistry courses, as well as courses in biochemistry and biotechnology. Fermentation Chemistry course (Washington University, St. Louis, MO).

EXPERIENCE	1/98–8/98—Engineering Trainee, Washington State Chemical Company, Federal Way Plant, WA.

Worked on the main ethylene-oxide production line monitoring performance of the fluidized catalyst bed reactor. Worked on plant tests investigating the effectiveness of different catalyst particle geometry on bed performance. Also worked in the higher olefin distillation unit. After initial training, was responsible for the computerized production reports from these units.

Was given a raise and worked over the summer, rather than returning to school.

5/97–8/97—Computer Operator, West Coast Federal Bank, Seattle, WA.

Fed data into mainframe computer; compiled, printed, and distributed standard reports. These reports needed to be completed and distributed on a very tight time schedule.

9/96–1/97—Plant Engineer Trainee, Whatabeer Brewing Company, St. Louis, MO.

Monitored fermentation processes that control process conditions and release the malt at the proper time.

5/96–8/96—Operated a house painting company on my own. Employed three other students.

5/95–8/95—Sales clerk and cashier, Happy Home Remodeling Warehouse, Seattle, WA.

AFFILIATIONS   American Institute of Chemical Engineers Student Affiliate University of Washington track team

REFERENCES   Available upon request.

An Experienced Chemist's Chronological Targeted Resume

Michael made a good choice in choosing the chronological resume format because it emphasizes his steady career advancement. However, this steady advance could lead to some questions during his employment interview aimed at understanding why Mike wants to change jobs. Mike may want to briefly deal with this issue in his resume cover letter, but usually it is best to wait and let it come up during the interview. However, Mike should be prepared with a realistic answer that does not reflect negatively on himself or his employer.

Mike's resume is also an example of a targeted resume in that he focuses on the pharmaceutical industry. His resume would appear stronger if there was a way that he could emphasize any experience in Food and Drug Administration-approved drug-testing protocols. This could also be done in a cover letter if he does not modify his resume. Note that while Mike has, of necessity, listed his work (day) telephone number, he has listed a personal, not a corporate e-mail address. He does not want to give the appearance of using his current employer's facilities or resources for job hunting. Having an e-mail address implies Mike is technologically up-to-date and suggests he could be using the Internet in his job-hunting efforts. If you are going to be job hunting, get a personal e-mail box through a commercial service.

Mike has a lot of experience, but kept his resume length to one page. He did this by omitting purely technical skills from his skills and by describing his accomplishments in general terms. He also left out any description of his accomplishments in graduate and undergraduate school. This is reasonable since his objective and recent experience are all in management, and graduate school was a long time ago. Although a two-page resume for a chemist of Mike's experience is not unreasonable and he could lengthen his resume by including more technical details, this would weaken his focus on management.

By using drug trade names and noting annual sales of commercialized products, Mike emphasized that he delivered value to his employer. The commercial drugs he was involved in and being the first to use combinatorial chemistry at Worldwide emphasize his innovation skills.

The brevity of this fictitious resume illustrates another principle; it is much easier to keep a resume brief when you are not emotionally involved with the accomplishments you are describing. It was much easier for me to invent Mike's resume and keep it short than it would be to edit my own resume down to one page. Emotional detachment helps in eliminating unnecessary detail and keeping the resume focused on your skills and abilities most related to your objective. A big advantage a resume-writing service or a knowledgeable friend editing your resume provides is this emotional detachment.

MICHAEL MANAGER

727 Elm Street
Chicago, IL 60036
Day (312) 555-7778   Evening (847) 555-2122
e-mail: mmanager@bol.com

OBJECTIVE	Seeking a research management position in the pharmaceutical industry in which I can use my technical, management, and leadership skills to bring new drugs to market more rapidly.
EDUCATION	Ph.D. Chemistry, New York City University, May 1978 B.S. Chemistry, Upstate College, May 1973

EXPERIENCE

4/96–present
Department Manager, Drug Delivery Technology Department, Worldwide Pharmaceuticals, Chicago, IL

7/91–4/96
Research Manager, Oncology Department, Worldwide Pharmaceuticals Set up combinatorial chemistry group in the department using MSI software. Was the first combinatorial group at Worldwide. While manager, increased the number of drug candidates in the pipeline tenfold due to combinatorial techniques and commercialized two new drugs for lung cancer chemotherapy, WORLDWIDER® and WORLDWODER®. Annual sales in excess of $500MM.

8/87–7/91
Team Leader, WONDERDRUG® Manufacturing Team. Worldwide Pharmaceuticals
Led the team that developed the commercial process for manufacture of this drug from benchtop synthesis through plant start-up. Worked with our own plant personnel and with a toll manufacturer. Awarded three U.S. patents. Annual sales $50MM.

3/83–8/87
Senior Research Chemist, Oncology Department, Worldwide Pharmaceuticals Synthesized drug candidates. Three progressed as far as animal testing. Awarded seven U.S. patents.

1/81–3/83
Research Chemist, Allergies Research Department, Worldwide Pharmaceuticals Synthesized candidates. One became ingredient of Breathease™ nasal spray. Also developed final formulation of Skinease™ eczema skin cream. Annual sales $30MM.

5/78–1/81
Chemist, Hometown Drug Company, Smalltown, NY Synthesized oligomeric ethers as candidates for skin care creams.

SKILLS
Experienced in managing departments of as many as 120 people and leading teams of 12–20.
Have used PC Leader Project Management Software to coordinate work team activities. Work productively with plant personnel to bring new products to market. Fast learner willing to take chances as indicated by my being the first to explore combinatorial chemistry at Worldwide and put it to work.

AFFILIATIONS
American Chemical Society   American Management Association

REFERENCES
Available upon request.

For an Experienced Chemist

While James is a younger chemist than Mike, he also has considerable industry experience. Unlike Mike, he has stayed at the bench. Consequently, his resume deals far more with his technical accomplishments. Besides purely technical accomplishments, James emphasizes his interactions with industrial customers. As a result, his resume is two pages long.

James' resume makes him an attractive candidate for chemical companies selling to the mining, industrial cleaning, and oil industries. However, he has not truly targeted these companies in his resume. For example, his "Objective" statement makes no mention of these industries. Instead, while his resume emphasizes his accomplishments in surfactant product development and technology, it also emphasizes his versatility.

JAMES A. TUFFWAY
9999 Hardscrabble Lane
North Houston, TX 77183-4747
Tel. 281-555-2724 (Day)   281-555-7777 (Night)

OBJECTIVE   Capitalize on my surfactant knowledge in a product development position in which I can utilize my teamwork and leadership skills.

ACCOMPLISHMENTS

- As team leader for Mining Chemicals, managed and headed development effort resulting in the MINFLOAT® series of ore flotation surfactants. These more than tripled Amalgamated's sales to the mining industry making us #1 in this business. This resulted in four U.S. and 16 international patents. Presented six papers at trade association conferences and wrote two papers published in trade journals.
- At Amaglamated, developed Wishy-Washy™ surfactant formulations for washing transportation vehicles: planes, trains, trucks, and cars. Product sales current over $50MM annually. Developed two of Intergalactic Chemicals' synthetic drilling fluid additives: EVIRO-OK® 7 and series, EVIRO-OK® 8. These products widely used in oil industry. Work resulted in two U.S. and 12 international patents. Presented two papers at trade association meetings.

SKILLS

- Skilled surfactant chemist. Knowledgeable in basic surfactant chemistry, synthetic procedures, properties measurement, and applications technology in the mining and oil industries and in vehicle washing.

- Use Timeline™ Project management software for my own work and to co-ordinate work of my team members.
- Maximize productivity of available resources by outsourcing work as appropriate. This includes two basic research projects on flotation and using a contract lab for routine surfactant screening tests.
- Excellent written and oral communicator. Frequently call on customers with sales representatives and present seminars. Often present papers at trade association meetings, write trade journal articles, and write first drafts of product bulletins.
- Excellent people skills. Work very well with others as team member and leader. Aggressive mentor of technicians. At Amalgamated, got two technicians promoted to sales and one transferred to information services department.
- Experienced in supervising technicians. Also have had three students work with me over the summer.

EXPERIENCE

9/95–present	Amalgamated Chemicals, Team Leader—Mining Chemicals. Lead seven-member team developing new ore flotation surfactants. Both coordinate work of others and work on my own surfactant projects. Work requires in-lab synthesis and applications testing, working with customers on their lab-scale applications testing, scale-up of product manufacture, and supervising full-scale customer field trials. Use statistical design of experiments in much of this work. Also coordinate environmental and toxicity testing of surfactants we planned to commercialize for the mining industry.
9/91–9/95	Amalgamated Chemicals—Staff Research Chemist —Mining Chemicals. Developed ore flotation surfactant formulations. This required laboratory scale synthesis and applications testing. Used statistical design of experiments to determine what nonionic surfactants to test and the test conditions to be used.
7/88–9/91	Intergalactic Chemicals—Senior Chemist—Drilling Fluids Group, Intergalactic Chemicals, Houston, TX. Developed synthetic drilling fluid additives designed to have improved environmental compatibility and reduced formation characteristics downhole for use on offshore oil platforms. Work required laboratory polymer synthesis and applications testing, working with toll manufacturers on scale-up of product syntheses, and supervising field tests on both land and offshore platforms.
5/85–7/88	Research Chemist—Central Research Department, Worldwide Chemicals, Newark, DE.

Synthesized novel monomers for use in polymerization reactions to make polyether copolymers. Also worked on metathesis of functionalized olefins. Both these projects required synthesis and evaluation of novel catalysts.

EDUCATION   Ph.D. Chemistry, University of California, Berkeley. Thesis title: "Ring-opening Polymerization Reactions of Substituted Cyclopentenes." Resulted in two presentations at national ACS meetings and three published papers.

B.S. Chemistry, University of California, Los Angeles. Attended summer school two years to complete 24 credit hour minor in chemical engineering. Senior Thesis title: "Synthesis of Highly Strained Oxetanes." Professor continued the work after I left with another student and eventually published paper on our work.

AFFILIATIONS

- American Chemical Society, Member, Secretary, Delaware Section in 1987
- Society of Petroleum Engineers, Member, also member of Drilling and Completion Fluids Technical Committee
- The Mineral Society, Member. Served on national meeting program committees for 1996–1998.

A Career-Changer's Resume

Scientists and engineers who change careers face many challenges in the job market. One of these is preparing a persuasive resume. Like all resumes, it must persuade the reader that the applicant has the necessary skills to do the job well. Due to lack of experience, it is more difficult for career changers to persuade resume readers of this. To do so, the career changer should gain some experience in the new field through part-time work, volunteer work, or professional society activities prior to looking for a full-time position. A solid record of previous accomplishment, even if in another field, will help persuade resume readers that you are a results-oriented individual who will succeed in your new field. However, any experience in your new field should be placed in a prominent location to help your resume survive initial screening.

Wendy Wordsmith is a chemical engineer who wants to become a technical writer. She wants to switch fields in part so that she can work more at home. However, it is unwise to bring this up until her employment interview. Instead, she has targeted her resume to companies producing or using chemicals and is emphasizing the value of her chemical background.

Looking at her resume, she wisely has not used the chronological format. Instead, she has used a skills format that emphasizes how much technical writing was a part of her previous job assignments. Note how she has divided her previous writing experience to increase the size of the "Accomplishments" section and impress the casual reader. She notes that she has written freelance articles. Having mentioned published articles, she should include a list of her publications giving the precise references so resume readers can locate and read the articles. Only after listing her writing-related accomplishments does she note her more-traditional chemical engineering accomplishments.

Wendy has also noted that she took a minor in English in undergraduate school. To emphasize this, she has placed her "Education" section prior to her work history. Her professional affiliations also indicate her interest in technical writing. If she had worked on the school newspaper, it would have been beneficial and she could have included this in a section on college activities including the word newspaper as a keyword for computer searches.

Wendy's "Objective" section may appear overly detailed, but provides a wealth of keywords for computer scanning. Her "Work History" section emphasizes her productive work on teams. This is important because corporate technical writers usually work closely with others; the people who provide her with writing assignments.

Wendy A. Wordsmith
911 Primrose Lane
Philadelphia, PA 24712

Tel: (216) 555-8234   e-mail: waword@bol.com

OBJECTIVE: Corporate position as a technical and business writer using my engineering background to write product brochures, advertising copy, press releases, material safety data sheets, plant operating manuals, and environmental impact statements.

ACCOMPLISHMENTS

- At Advanced Chemicals Co., I wrote the plant operating manual for the production of three pharmaceutical intermediates: ADCHEM 327, ADCHEM 329, and ADCHEM 330.
- Also wrote product bulletins on these three products and the first draft of the material safety data sheets for these products.
- Written two articles published in *Pharmaceutical Industry Monthly*.
- Have written six freelance articles published in consumer and business magazines.
- Optimized the base-catalyzed, stereospecific bimolecular elimination process

step for the manufacture of ADCHEM 327, ADCHEM 329, and ADCHEM 330 in laboratory and pilot scale studies. Supervised the first two runs of this process in the commercial plant.

• Awarded a U.S. patent for the base-catalyzed, stereospecific bimolecular elimination process step making the common intermediate for the manufacture of the three ADCHEM products listed above.

EDUCATION

• M.S. Chemical Engineering, February 1996, Ohio State University, Columbus, OH. GPA 4.0
• B.S. Chemical Engineering, May 1993, University of Rochester, Rochester, NY. GPA 3.47
• Minor in English taking seven courses (21 credit hours) including Technical Writing and Business Writing courses

WORK HISTORY

2/97–present   Research Engineer, Intermediates Department, Advanced Chemical Company, Philadelphia, PA. Member of ADCHEM pharmaceutical intermediates process development team. My project was development of a process step to make a 100% cis-olefin product which could be used as a common intermediate for the synthesis of compounds of interest to the pharmaceutical industry. Three of these compounds were commercialized and are currently sold to pharmaceutical companies. The ADCHEM Team won the 1998 Outstanding Team Performance Award that included a $5,000 bonus for each team member.

2/96–2/97   Research Engineer, rotating asignment, Advanced Chemical Company, Philadelphia, PA. The standard initial assignment for new hires is to work in several departments on temporary assignment. I worked four months in two assignments in the Rockville, MD, plant, three months in the Gainesville, FL, plant, two months in the Intermediates Department, two months in the Agricultural Chemicals Department, and one month in the Distillations/Purification Department.

AFFILIATIONS   Society of Technical Writers
                  American Institute of Chemical Engineers

REFERENCES   Available upon request

14

Resume cover letters

The content and the format of your resume may enable you to survive the initial employment screening process. However, when the hiring authority sits down with the surviving resumes, cover letters play a large role in deciding who gets employment interviews. Your cover letter is at least as important as your resume, according to 60% of the large company executives surveyed in 1995 by Accountemps, a temporary staffing service located in Menlo Park, California. Even with the other 40%, a carefully crafted cover letter can only improve your chances to obtain an employment interview.

Cover letters serve to tailor your sales package to a particular job opening. Thus your cover letter should identify the job opening for which you are applying or the types of jobs for which you are best qualified. This is particularly important if you send your letter to a personnel manager or a recruiter. Your cover letter can also help the hiring authority begin to answer the very important question "Will I like this person". Your personality, especially enthusiasm and creativity, can come through in a way that just is not possible via the resume. Your cover letter, if mailed to recruiters, can give them a sense of your personality and goals. This helps them focus on the most appropriate job openings for you. The personal impression your letter provides helps them to better describe you to their client companies.

There are several steps you can take to write a powerful cover letter that will impress prospective employers. The best way to write a cover letter is not to start at the beginning and work your way to the end. Before you start writing, you need a plan.

PLANNING

Begin by doing your homework. Find out what you can about the employer in general and the available job opening in particular. Review your qualifications and accomplishments for features that will particularly interest this employer. To do this, pretend that you are the employer. Ask yourself what abilities or special qualities a job candidate needs for this job opening. Then review your own experience and accomplishments. Include job assignments and special projects. Recent students should include courses they have taken, special projects, extracurricular activities, and their jobs. Consider how these demonstrate that you have the abilities and personal qualities needed for the job.

Write down examples of your accomplishments and abilities. Make them more than a list of facts. The most common error made in writing cover letters is to duplicate information already in your resume. Take a fact listed in your resume, expand it, and put your personal imprint on it so that your cover letter example tells a short story. Use each example to illustrate that you have the qualities the ideal job candidate needs.

WRITING THE BODY OF THE LETTER

Your examples will become the body of your resume cover letter. Focus on your most important two or three examples. These may change from one cover letter to another as you tailor your letters for specific employers or job openings. Discussing more than three in your cover letter will blur your focus and lengthen your resume cover letter too much. (Save the other examples and mention them in your employment interview later.) Chemists, engineers, and technicians increasingly work in teams. Look at your experience and qualifications to find examples that demonstrate that you are a team player and work well in groups.

No job candidate is ideal. In writing your examples, emphasize the positive without exaggerating. Realize the difference between accomplishments and benefits. Benefits are the outcome of what you accomplish. Presenting a paper at a trade association meeting is an accomplishment. Being able to say, "My conference presentation resulted in customer inquiries that produced three million dollars in first year sales" is a statement of both accomplishments and benefits. Be sure to avoid disclosing proprietary information about your current or former employer when describing benefits. However, quantify benefits when you can; numbers can impress readers.

Focus on what you can do for the employer, not on what the employer can do for you. Chemists and engineers are problem solvers, as are technicians. Emphasize how you can contribute and help the employer solve particular problems. Discuss how your abilities match the job requirements, not on how the job opening matches your interests.

Use appropriate language. Employ industry-specific jargon you picked up from trade journals. Be certain to use any jargon correctly. However, do not overdo the use of jargon or your tactic will become obvious.

THE LETTER OPENING

You should now have a draft of the body of your resume cover letter. Next comes the most critical step in preparing your letter: writing an attention-getting introduction. Your cover letter is a sales letter. People seldom continue reading sales letters if the opening does not grab their interest. Sales people call this "having a handle." The handle is the special attraction that you have for a particular employer.

This first sentence must grab the reader's interest immediately. You can do this in several ways. The first is to offer an immediate benefit to the employer. The engineer who can say, "I invented a process that saved my employer three million dollars a year" commands attention. So can the chemist who can open his cover letter by saying, "My knowledge of the mining industry chemicals can help XYZ Company increase its sales to this highly competitive market." However, avoid coming on too strong since your readers are mostly conservative types. For example, consider the arrogance implied in, "If you want to succeed in the mining chemicals business, you need to hire a chemist like me."

One cover letter tactic professionals seldom use is opening with a question. For example, "Do you want to hire a chemist whose twelve food additive patents earn more than twenty million dollars in annual sales?" Again, avoid appearing conceited. "Do you want to hire a chemist who can guarantee XYZ increased profits from your ore flotation chemicals business?" is an example of confidence crossing the line into arrogance.

Referring to recent news about the company can be an effective first sentence in a cover letter. For example, "XYZ's recent acquisition of WOW Chemicals means your sales to the mining industry could grow rapidly." Follow this with, "My mining chemicals experience can help exploit this new opportunity" and your readers will probably be hooked.

Name dropping can be helpful. Mention the names of any contacts you have in the company if they can provide favorable references or had notified you of the job opening. Focus on those who are doing well. If a professor who is widely recognized as a leading expert in his field recommends that you apply to a particular company or for a particular position, this would be another good name to use. Before you use anyone's name, be sure to get permission. Name dropping in a first sentence can be an effective opening if you pick a powerful name. Take advantage of your networking. Suppose John Smith is an eminent company researcher or senior manager. Writing "Mr. John Smith mentioned to me that XYZ Company is looking for qualified polymer chemists and suggested I contact you" can be a powerful opening. It is best if the person you cite is able to give you a favorable recommendation.

This recommendation does not have to be in an on-the-job context. Suppose you served on a conference-organizing committee and did an excellent job. The chairperson, Dr. Ruth Smith works for XYZ Industries. As a job hunter, contact Dr. Smith to inquire about opportunities at XYZ. If appropriate, begin your letter with the statement, "Dr. Ruth Smith of your Fermentation Chemicals Department mentioned XYZ Industries' need for a qualified carbohydrate chemist and suggested I apply for this position." If asked, Dr. Smith can say you are a hard worker and comment favorably on your organizational skills. Even if she cannot comment on your technical skills, you have a strong cover letter opening.

Use chance encounters to your advantage. For example, I once sat next to John Bookout, former chief executive office of Shell Oil Company, on a long airplane flight. We discussed detergent chemicals and their use in various industrial applications. If I had been job hunting and wrote Shell Oil Company, I could open my cover letter by saying, "While discussing detergent chemicals with John Bookout, I became convinced that Shell could benefit from my expertise in this area." I would then go on to describe how Shell could benefit. You can also use this approach in a less-aggressive lower-key context. An example of this approach is, "Knowing of my interest in employment opportunities at XYZ, Dr. John Smith suggested I write you."

Many applicants begin cover letters by referring to the employer's job opening advertisement. While unoriginal, this opening immediately identifies the job for which you are applying. Another low-key and frequently used first sentence is a description of your career objective. Tailor this to the job opening. However, some experts feel stating a career objective implies a focus on what the employer can do for you rather than how you can help the employer. Remember, the reader wants to hire someone who can benefit the company and make the hiring authority look good.

An introduction should be short, but more than just one sentence. When writing the rest of the introduction, again, put yourself in the employer's place. They do not want to solve mysteries; they want to spend as little time as possible finding worthwhile employment candidates to interview. Be sure you tell the reader exactly what job you are applying for. Any confusion and the reader will quickly lose interest.

CLOSING THE LETTER

End your letter decisively. Next to the introduction, the closing is the most important part of the cover letter. Do not let your letter meander to an indefinite or weak close. A strong close projects an image of you as an aggressive and decisive person.

Usually, the best way to close is by asking for an interview. Provide your home phone number so employers can contact you quickly once they decide to interview you. Inform the employer that you will call in 10 to 14 days. Be specific about the date and call exactly on schedule. The best times to call are before 8:30 A.M. and after 5:00 P.M. as the person you are calling is less likely to be in meetings at these times. Do not let fear of rejection prevent you from making this call. Use this call to verify receipt of your letter, to learn about the process being used to fill the job opening, and to ask for an interview.

AVOID WRITING MISTAKES

We already mentioned two mistakes to avoid: simply repeating information in your resume and focusing on your own goals and requirements rather than on how you can help employers meet their objectives. A focus on yourself is often indicated by many sentences beginning with "I."

There are several other mistakes to avoid in your cover letter. It is important to avoid sexist language in your letter. Do not assume your reader is a man. Avoid the salutation, "Gentlemen." Find out the name of the appropriate individual, if possible the hiring authority, and address the letter to this person. Avoid negative subjects. For instance, do not mention being out of work or how badly you want to change jobs. These subjects are difficult to discuss in writing. Mentioning them in your cover letter can cost you a job interview despite your excellent resume.

Do not waste words on the obvious. For example, "During my senior year I worked at XYZ to earn money." The last three words are clearly unnecessary. Perhaps the most common example of this is the opening sentence, "Enclosed is a copy of my resume." Your reader already knows your resume is there. Avoid mentioning your salary history or salary requirements. These are matters for later discussion. The same is true for geographic restrictions. These have the effect of restricting employers' interest in you. Also, the right job opportunity or a lengthening period of unemployment may lead you to rethink these requirements. You do not want to be put into a position of having to retract or modify them. This makes you appear indecisive.

Excessive formality and overdone courtesy are signs of insecurity. By all means, be courteous and business-like. However, wordy phrases such as "Thanking you in advance" and "In anticipation of a favorable reply" suggest inexperience and also drain a letter of its personal flavor. Phrases such as "If my qualifications are of interest, please contact me at your convenience," make you appear weak and passive. While projecting personality is important, do not go too far in the opposite direction. Your cover letter is a business document, not a note to a friend.

There are also several aspects of letter style and mechanics that can doom an otherwise well-written letter to the wastebasket. These are:

- Impersonal salutation. Always open with a person's name, not "Dear Sir," "Dear Madam," "Dear Research Manager," and so on. Consult placement offices or professional and trade association membership directories for names of appropriate contacts. Thanks to personal computers, you can begin keeping a database of these before you begin mailing out your letters and resumes. Students may find such a database an excellent activity for an ACS student affiliate chapter. However, the key is to keep this database continuously updated.
- A mass-produced appearance. You want your cover letter to appear tai-

lored to the employer and written just for her. An impersonal salutation can give your letter a mass-produced, impersonal look, as can never mentioning the employer's name other than in the address. Citing examples of your accomplishments that are inappropriate to the employer can also doom your resume and cover letter to the wastebasket. Word processors make it much easier to personalize letters.

- Getting too informal and personal. There are some things better left unsaid. For example, do not write, "I have been researching career opportunities that would let me relocate to Minnesota and be near my parents." While such comments might not concern most readers, they will bother some. After all, what happens if the parents retire to Florida? Also, while you want your personality to show in the letter, remember that this is a business letter.
- Inappropriate letter length. A very short letter that just says, "Here's my resume" implies haste and carelessness. An overly long letter burying the reader in many details implies poor judgment.
- Poor writing. A letter with poorly written sentences or a poor flow of ideas demonstrates inadequate written communications skills. Would you want this person writing letters to your customers?
- Poor letter mechanics. Many readers will throw out a cover letter or résumé as soon as they spot a typo or misspelling. With word processors widely available, there is no excuse for these. Similarly, word processors make proper formatting easy.

EDITING YOUR LETTER

Edit your cover letter carefully. It must be easy to read. Focus and clarity of expression in your letter imply focus and clarity of thought, which are very desirable qualities in an employee. Make both your cover letter and resume be examples of your best writing. Use action verbs and avoid the passive voice. Sentences should be fairly short. Check grammar carefully. If you are using a word processor, take advantage of the spell and grammar check functions, but do not rely solely on these checks. Proofread your letter carefully.

Your cover letter should not be more than one page long. The critical question is whether your personality, qualifications, and accomplishments meet the employer's requirements. Anything else should be ruthlessly edited. Paragraphs should be short with a blank line between them. Large blocks of text intimidate the reader.

If you have the time, set the letter aside for a day or two before editing it. You will be more detached and can more easily edit the letter, especially if it needs to be shortened. The flow that seemed so obvious when you are first writing may seem much less so a day later. Typos and spelling errors will also be easier to spot. Check to be sure the business address is correct and you have spelled the person's name correctly and included the proper title—Dr., Mr., Ms., etc. Carefully review both the cover letter and your resume. Be sure information in them should be completely consistent.

Print your letter on good-quality white or ivory bond paper. Left- and right-hand margins should be at least one inch, as should the bottom margin. The date at the top of your letter should be at least two inches from the upper edge of the sheet. Margins give the reader a place to write short notes.

The most common lament of job hunters is the limited response they usually receive to a mass mailing of cover letters and resumes to prospective employers. Because of the time required to just add an address to a form letter, many companies do not mail rejection letters to job hunters. Most career counselors believe mass mailings of cover letters and resumes account for just 5–10% of all jobs filled. For jobs where technical skills are important, direct mail contact may account for somewhat more of the jobs filled. These jobs include scientific research, engineering, information science, and computer specialists jobs. If you are unemployed, you cannot afford to ignore the mass mailing option, but use it as only one of your job-hunting methods. If you are job hunting while still employed, mass mailing can compromise the confidentiality of your job search and a combination of other methods is usually more appropriate. You need to avoid the mass mail look by targeting your resume and cover letter to specific industries, if not specific companies. Also address your mailing to a certain individual, not an anonymous research manager. If possible, focus on people with hiring authority. Try to avoid mailing your resume and cover letter to the personnel or human resources department. However, be prepared to make exceptions. For example, if you can identify the name of the person in charge of technical recruiting, it may be valuable to send her your resume and cover letter. She can route your materials to appropriate departments that may have job openings. This could result in an employment opportunity whereas sending your letter and resume to a single hiring authority does not. For example, consider a chemist with experience in water-soluble polymers. A polymer department research manager might discard this resume because she has no job openings. A manager of technical recruiting at the same company would forward it to the manager of water-treatment chemical development with a job opening available. Since some water-soluble polymers are used in water treatment, our job hunter could be qualified for this job opening and receive an employment interview.

There are several tactics to identify the hiring authorities and avoid being forced to send your letter only to a human resources department. You can use your network; contact people you know who work for a company and ask them for names of the appropriate hiring authorities. They may be able to tell you which departments have openings and what qualifications they are looking for in an applicant.

You can use professional society membership directories. Some organizations, in addition to listing members alphabetically, include a section in which members are listed by their employer. You could look up XYZ Company and see that 22 of their employees are members of a professional organization in your specialty. If job titles are included in the listings, you could then determine which are the managers who are likely to have hiring authority.

One critical issue is how to make your resume and cover letter stand out from the many others the company receives. Target your letters and resumes for specific industries, or perhaps have a mentor write your cover letter (see Appendix to this chapter). Contact smaller employers to reduce the number of competitors. However, be aware that many smaller companies prefer to find employment candidates nearby to avoid travel expenses and later relocation costs when they hire someone.

FINAL REMARKS

An excellent cover letter can make the person described in a resume come to life. This can only increase the prospective employer's interest and improve your chances of making it to the next step of the job hunting process—the employment interview.

ADDITIONAL READING

Developing the Right Picture: Résumé Preparations. Video. American Chemical Society: Department of Career Services, Washington, DC (1998).

Formula For Success: Turning Job Leads into Gold. Video. American Chemical Society: Department of Career Services, Washington, DC (1998).

Beatty, R. H. *The Perfect Cover Letter.* Wiley: New York, 1996.

Besson, T. *National Business Employment Weekly Cover Letters.* Wiley: New York, 1996.

Hansen, K. *Dynamic Cover Letters.* Ten Speed: Berkeley, CA, 1995.

Ireland, S. *The Complete Idiot's Guide to the Perfect Cover Letter.* Macmillan General Reference: New York, 1997.

Marler, P. *Cover Letters Made Easy.* NTC Contemporary: Lincolnwood, IL, 1995.

Yate, M. J. *Cover Letters That Knock 'Em Dead,* 3rd ed. Adams, Adams Media: Holbrook, MA, 1997.

For Midcareer Job Hunters

Enlow, W. S. *201 Winning Cover Letters for $100,000+ Jobs: Cover Letters That Can Change Your Life.* Impact: San Luis Obispo, CA, 1998.

APPENDIX

The Sponsor's Cover Letter

Your resume and cover letter are likely to be in a stack with many others—some from applicants at least as qualified as you. Can you write a cover letter that will really make you stand out from the other applicants? The answer is that you do not write the cover letter at all! Instead, ask a mentor to do so. This personalizes your letter and distinguishes you from your competition. Your cover letter also becomes a recommendation.

These letters work best for graduating students and post-docs. For someone with even a limited amount of industry experience, using a mentor to write your letter implies a lack of independence and self-confidence. A mentor's letter can help a job hunter overcome the disadvantage of limited networking.

Begin by choosing your mentors. They should know you well and be willing and able to attest to your abilities. In other words, these are the same people you would choose as references (see chapter 15). For students, professors who know you well are the obvious choices. Particularly valuable are professors with whom you have a one-on-one relationship, such as faculty advisors. If you undertook a special project, such as undergraduate research for a professor, this person would be a good mentor and letter writer. Students' supervisors on part-time and summer jobs also make good mentors. However, the more different the job and the position you are applying for, the less credibility your supervisor will have. For instance, suppose you are a chemical engineering major and worked as a waitress on a summer job. The restaurant manager can attest to your personal attributes such as teamwork skills and enthusiasm, but not to your engineering capabilities.

◇ Sponsor's Letter to a Personal Acquaintance

Mr. Earl Employed, Manager, Polymers Development Department
Polymers Department
Really Big Chemical Company
1111 Mill Way
Mill City, California 77777

Dear Earl:

One of my students, Florence Flask, has just finished her Ph.D. degree and is looking for an industrial R&D job. She completed the Ph.D. in less than four years. Florence did her thesis research for me on the electrical properties of dendritic copolymers of 3-methylthiophene. Therefore, I know first hand that Florence is a very capable chemist and a take-charge person. She is mature for her age and works well with others of diverse cultural backgrounds.

If you are hiring research chemists, I can heartily recommend Florence. If you do not have any openings, I would appreciate if you could call Florence at (555) 555-8888 and give her input on possible job opportunities elsewhere and employers' needs from your perspective.

I am attaching Florence's resume. Should you want to learn more about her capabilities and accomplishments, by all means, give me or Florence a call. My office number is 555-918-1234.

I hope this letter finds you, Marge, and the kids well. I hope to see you at the next ACS meeting.

Sincerely yours,
Dr. Peter Professor
Department of Chemistry
First Rate University
College City, New York 11111

Once you have picked your mentors, the next step is to pick the targets, both companies and individuals, for your mailing. The place to start is each mentor's Rolodex. Your mentor's letter will be more persuasive when sent to someone in her network than when sent to a stranger.

You should also assemble a second list of people and companies your mentor does not know. Sources for these names include industry listings, membership directories of professional societies and trade associations, job advertisements, and the telephone book. This approach will personalize your cover letter when it is sent it to someone your mentor knows. However, even when your mentor is writing to a stranger, your cover letter has a unique personal touch. The letter says that your mentor is so impressed by your abilities that he is willing to say so in print. This personal touch also helps to avoid your letter looking mass-produced.

Approach your mentor with the concept. If he agrees, help him write the letter or write it yourself—whatever he prefers. Follow the guidelines to write a conventional cover letter. The letter should reflect your mentor's communication style. If your mentor is emphatic in stating opinions, the cover letter should reflect this without being abrasive. If your mentor is low key, the letter should reflect this while still being supportive and persuasive.

Your mentor must see the final version of the letter. It must be something she is comfortable with—honest in content and having a tone consistent with her personality. Ask to have the number of copies you need printed on a letter-quality printer using the stationery of the mentor's employer. Try to get your mentor to sign all the cover letters. If the letters are going to people the professor knows, you should certainly get him to sign them. Also, if your mentor should later write a reference for you, his signature should match that on your cover letter—another reason to get him to sign all copies of your cover letter. Only if you cannot get your mentor to sign the cover letters himself should you ask permission to sign for him. Avoid using a rubber stamp signature. This will give your letter a mass produced appearance. (Another approach is for your mentor to send your resume and her cover letter to colleagues she knows via e-mail.)

Even if you have a long list of companies to contact, you should mail your letters and resumes in small batches of 10–15. The small batch size will make your mentor more willing to sign the cover letters himself. Get your most important letters out sooner by prioritizing the mailing on the basis of companies you would most like to work for or company location. Follow up your letters by telephone. The task is more manageable when you have only 10–15 calls to make per week. The same week you are making follow-up calls, you can also be sending out additional resumes and cover letters to companies lower on your priority list.

Sometimes a lack of response from prospective employers can indicate you are not selling yourself effectively. Switching your approach and contacting these companies again with a mentor's cover letter can add fresh life to your personal marketing campaign. (When doing this, you should also consider rewriting your resume. While changing emphasis is acceptable, think carefully before making a major alteration such as changing your career objective. Such

---

◆ Sponsor's Letter to a Stranger

Dr. Kenneth Kinetics, Technical Superintendant
World Scale Chemical Plant
Colossal Chemical Company
9999 Winding Way
Chemistry City, California 88888

Dear Dr. Kinetics:

I am writing to professional colleagues to introduce Ben Beaker. Ben has worked with me for four years as a chemical technician. He has always done an excellent job and I will be sorry to lose him. However, Ben has decided that he wants to move from the lab to a chemical plant and no suitable positions are available with our company. I strongly recommend Ben to you and am enclosing his resume.

Ben has worked with me on various process development projects for four years—both in the lab and in our pilot plant. He has also worked for weeks in our

Louisiana plant on process start-ups. Therefore, I know first hand that Ben is very capable and familiar with plant environments. He has developed excellent teamwork skills. I believe he can do an excellent job for Colossal in your plant. He is definitely capable of becoming a plant chemist or unit manager.

I am enclosing Ben's resume. Should you want to learn more about his capabilities and accomplishments, please call Ben at (555) 555-1111 or me at (555) 555-0007.

Ben and I appreciate whatever help you can provide. I hope I can return the favor in the future. I look forward to hearing from you.

Sincerely yours,
Dr. Edward Erlenmeyer, Sr. Research Chemist
Research Center
Detergent Chemicals Company
Reallyclean Town, Texas 66666

changes could present an image of indecision to prospective employers who have kept your earlier resume on file.)

Your mentors have done you a big favor by signing your cover letters, and you are obligated to them. Keep them informed of job leads, interviews, and job offers you receive. This will make them feel good and see the help they gave you as a win-win situation.

15

WORKING WITH YOUR REFERENCES

Good references are a key factor in a successful job search. How can you properly select your references and reliably predict what they will say about you? Just as you tailor your resume to a particular job opening, so, too, must you tailor the list of references you provide each employer. Ideal references are individuals who can serve as strong advocates. They must be able to clearly describe your skills and accomplishments'. They must be able to focus on your accomplishments most relevant to the job opening. They should be articulate and pleasant in manner. Your references should be willing, even eager, to help you land a new job.

Employers will vary in how they consult your references. Some will want to talk to them before your employment interview. Others will contact them after the interview and before making a formal job offer to you. And some firms will not contact your references at all. Some personnel department managers will only contact your references to verify your education and dates of employment. Others, usually hiring authorities, will question references in depth. They will use this information to assess your strengths, weaknesses, and personality traits. A prospective employee's growth potential and adaptability to changing priorities and job assignments are of critical importance in today's reengineered corporations. Most companies are very interested in quizzing your references to see if you have these desirable attributes.

CHOOSING YOUR REFERENCES

Job hunters should not wait for employers to request references. Choosing references should begin when you start your job hunt. Identifying individuals to serve as references should be part of assembling information for your resume and stating your job objective. With each piece of information, list those people who can discuss this aspect of your career. At this stage, you are brainstorming. You are trying to keep your mind open to new ideas, do not try to eliminate names from your preliminary list at this stage.

Once you have your resume drafted and a job objective defined, it is time to select references from your preliminary list. Choose those who can provide information most pertinent to your job objective. Be sure these individuals can clearly describe the relevant aspects of your abilities and accomplishments. (Do not worry if their information will be out of date at this stage. As discussed below, you will be contacting them.)

For example, consider Barry Bench, a midcareer bench chemist looking for a staff research position in polymer synthesis. His references should be able to describe Barry's abilities and previous accomplishments relevant to this goal. These could include monomer synthesis, work in various areas of polymer chemistry, ability to supervise technicians, teamwork skills, and communications abilities.

Suppose you are looking for two different types of positions. You will need separate resumes and perhaps different references for each position. Consider the graduating chemistry student Nancy Novice. She is looking for a job in R&D or as a plant chemist, or in chemical sales. References for the first two positions should be able to attest to Nancy's laboratory skills and technical problem solving abilities. References for the chemical sales position must be able to describe her oral communications skills and her ability to work with people.

What if you are planning to make a career change? You will need to identify references who can discuss your abilities most applicable to your new field of endeavor. For example, consider bench chemist Andy Atom. Andy wants to leave the bench and find a job as a technical writer. While he has done some technical writing for his current employer, there are no technical writing positions available within his company. Andy wants to keep his job search confidential, so he is naturally reluctant to use coworkers as references. What can he do? If Andy is a recent graduate and did an excellent job when he wrote his thesis, his thesis advisor can serve as a reference. If he worked with editors while he wrote technical papers and review articles, they may provide good references. Andy's published papers can serve as testimony to his writing skills, as well as his technical abilities. What if Andy fails in finding more than one person who can testify to his writing skills? He has learned an important lesson: he probably needs additional preparation (see chapter 9) before he can realistically consider changing careers.

If you are job hunting while still employed, you have to rely on your references' discretion in your job-hunt confidential. If you have any doubts about their discretion, do not use these individuals as references.

INTERACTING WITH YOUR REFERENCES

Once you have your list of references, ask permission to use their names in your job search. Stress their value to you. Make them feel like partners in your job hunt. Do not take their help for granted. They are providing a valuable service, so you should treat them with dignity and respect.

When someone agrees to serve as your reference, describe the type of position you want and explain your reasons for the job change. Give them a copy of your resume. As your job search proceeds, inform them of your progress. Many references will be eager to offer advice and encouragement.

Opinions vary about when job hunters should provide references' names to employers. It is best not to list their names in your resume or cover letter. Instead, note that you will provide references' names upon request. This will keep you in control of the process as long as possible. Also, when the employer calls to ask you for names of references, you can learn more about the job opening. This will help you choose the most appropriate references for this employer to consult. Then, call your references and provide them with updated information about the position. This update will help to emphasize your most relevant abilities and accomplishments. Just by alerting your references to expect the employer's telephone call shortly, can help them better prepare to discuss your abilities.

Midcareer chemists and students face different problems in choosing people to serve as references.

MIDCAREER DILEMMAS

Job turnover is now so high that midcareer chemists often lose contact with former supervisors and colleagues. This is unfortunate since these individuals could serve as excellent references. Continuous networking can minimize this problem. Some professional organizations publish annual members' directories; utilizing this tool and calling an organization to get a member's telephone number and address can help you contact colleagues with whom you have lost touch. The ACS Directory of Graduate Research and College Chemistry Faculties can help you locate former professors who have changed jobs.

Many employers have policies that prohibit discussion of former employees other than verifying dates of employment and positions held. Continuous networking and professional society activities can add to your reference list and overcome this problem. In particular, trade association activities allow you to demonstrate your technical abilities, knowledge of industry problems, and your teamwork skills to people outside your company. These people may work for prospective employers or could serve as references. Colleagues who can state that you delivered several excellent papers at trade association meetings are valuable references, as well as those who can testify to your teamwork skills in serving as an officer, putting together meeting programs, and working on special projects.

Professional society activities and industry recognition may mean that colleagues working for your prospective employer know you well enough to serve as references. This sounds attractive; however, caution is necessary if you ask these individuals to serve as references. You need to be certain that these colleagues are comfortable with serving as your references. Only ask those individuals you are certain are highly regarded by your prospective employer. These individuals should like and trust you since their credibility is on the line more than that of an outside reference. They need to be familiar with your work and accomplishments. Vague supportive statements, even if they come from within the prospective employer's organization, are not very helpful. This is particularly true if the prospective employer contacts these individuals, rather than people outside their organization who can discuss your abilities and accomplishments in more definite terms.

There is an additional factor to consider. Many employers post job openings internally to try to fill them with current employees. Only if this tactic fails will they go outside the company to hire someone new. Suppose an individual at your prospective company is under consideration for the same job opening or was previously rejected. In these circumstances, you may be unlikely to obtain a helpful reference from this person.

DEFUSING NEGATIVE REFERENCES

Negative references from former employers can be a major problem. A negative reference can have a very adverse effect on a job hunter's success. Make every effort to avoid them. What can our midcareer chemist, Barry Bench, do about this problem? We will assume that Barry is a competent chemist who made some mistakes in handling his work relationship with a difficult supervisor.

Barry can employ four possible strategies. First, he can try to repair a poor relationship with his former supervisor. He can call or write to ask her to serve as a reference. Barry should state his request positively and in a friendly manner. He should note that one reason he is contacting his former supervisor is the professional respect he has for this person. This is a touchy situation, so Barry should script any conversation in advance. By listing key points, Barry can keep the conversation focused on his current situation and not on subjects that will raise old negative feelings. In any discussion of former problems, Barry should emphasize how much he has learned since the problem period. He should strongly state that he has become more skilled in working with others and would never allow such problems to recur in future jobs.

If Barry writes, to his former supervisor before telephoning, he should have a trusted colleague review his letter. This colleague should have excellent written communication skills. This review will help assure that the former supervisor cannot misinterpret Barry's written statements in a negative way. Unless his former supervisor is hostile, Barry should follow up any telephone conversation with a letter. He should thank the supervisor for her time and summarize the key points of the conversation. Barry should enclose a copy of his resume. This is particularly important if Barry has made significant accomplishments during the period after he worked for this supervisor. He may find that time has moderated the former supervisor's negative feelings. If Barry is unemployed, the supervisor may feel guilty about this. She may even be a victim of corporate downsizing herself. As a result, she may be sympathetic to Barry. These changed feelings, plus an agreeable and mature attitude on Barry's part during the telephone conversation, substantially improve the reference from this former supervisor.

Barry has other damage-control options even if he and his former supervisor now work at different companies or if Barry is unemployed. Suppose his former supervisor is active in a professional organization or civic group. Barry could join this group and try to work on a joint project with the former supervisor. He should be completely honest and inform his former supervisor he wants to improve her opinion of his interpersonal skills and technical abilities. He should explain that he wants to work with her in a different context to do this. A successful project and a productive, if not cordial, new work relationship can greatly improve this former supervisor's reference. Barry will have distinguished himself as a professional who can learn from mistakes.

If Barry is employed and only contemplating job hunting, he can take other steps to improve his relationship with a former supervisor employed at the same company. He can visit her and admit he made mistakes in handling their work relationship and would like to repair their relationship. An excellent way to do this would be for Barry to work on an appropriate project with his former supervisor. Initially, the working atmosphere may be strained. However, if they persevere, Barry and his former supervisor can develop a more productive work relationship.

A second strategy is to rely on special circumstances to eliminate the negative individual as a possible reference. Suppose Barry only worked with his former supervisor for a short time. This could be a valid reason for a prospective employer not to contact that individual. Barry could rely on his former employer's policy against giving out any information other than employment dates and positions held. In this case, he should try to provide alternative references such as a retiree from his former employer. This should be a senior colleague, manager, or more senior manager to whom Barry's supervisor reported. This individual should be familiar with Barry's work and accomplishments and willing to discuss Barry with a prospective employer. If this reference knows about Barry's problems with his supervisor, Barry should discuss the situation with him. He should acknowledge where he was at fault and stress that he will not allow a similar situation to develop in the future. Barry should not ask the reference to lie. However, he should request that the reference only comment on his work relationship with the former supervisor if a prospective employer asks. Barry's honest discussion and constructive attitude could persuade his reference to minimize Barry's past problems with a former supervisor.

A third strategy is for Barry to tell the prospective employer that he has had excellent work relationships with all of his former supervisors except one. He should provide the names and telephone numbers of these other supervisors. Finally, he should describe his relationship with the former supervisor briefly, honestly, and unemotionally. Above all, Barry should avoid bitterness, anger, and exaggeration. The key is to be constructive. Barry should emphasize what he learned from the situation and how he will prevent similar situations from recurring in the future. He should only volunteer this information if asked to provide the name of this former supervisor.

The fourth strategy is risky. That is to do nothing. This may be a viable strategy if Barry knows that his former supervisor is often out of the office and neglects to return phone calls. However, a vindictive former supervisor could depart from character, return the call, and provide a negative reference.

STUDENTS' REFERENCE OPTIONS

Students often have a limited choice of references. Because of this, some students list professors who barely remember them. These professors make poor references because they often give vague responses to employers' questions. In tight job markets, this can kill a student's employment chances. As the time approaches to job hunt, students should try to develop closer relationships with at least two professors. Working on special research projects for a professor is a good way to get to know a professor well and obtain an excellent reference, as is becoming active in a student chapter of a professional organization and getting to know the faculty advisor.

Be creative. Contact the former chemistry graduate student who was one of your lab instructors and is familiar with your laboratory skills and technical abilities. Yes, he may be a professor at another university or have another job now, however, he may still remember you. He will probably be flattered when you ask him to serve as a reference. Ask his thesis advisor or the university alumni association for his current phone number.

Consider going further afield for a reference. For example, suppose Susan Student is applying for a position that will mean writing a lot of technical service reports. A journalism or English professor who can attest to her writing skills could be a valuable reference. (However, she still needs at least one reference who can favorably describe her laboratory skills and chemistry accomplishments.)

Think about those individuals who can describe your interpersonal skills and organizational abilities. These could be supervisors or senior coworkers at summer or part-time jobs. Were you active in a student professional group such as an ACS Student Affiliate chapter? If so, the group's faculty advisor may be a valuable reference. These references are most useful when you are applying for nonlaboratory positions.

Also the unexpected. For instance, I did not realize for years that one of my undergraduate chemistry professors would be an excellent reference. He taught a large lecture course and I did not believe I stood out from the crowd. I was wrong. In addition, he had been getting periodic reports of my graduate school progress. As a result, he had a very positive opinion of me. However, because of my own limited thinking, I never capitalized on this during my job hunt. Only after several enjoyable visits with him at ACS national meeting alumni hours did I realize my lost opportunity.

REFERENCES FROM LEFT FIELD

Occasionally, prospective employers will contact someone they know who has some familiarity with your work. This is often someone you never considered as a reference. They often trust these "references from left field" more than those you suggest. This is because many employers place little value on staged, carefully scripted descriptions of candidates' accomplishments and abilities, particularly from people the hiring authority does not know.

Sometimes you can exert a bit of control over this situation. For example, consider Susan Student. She knows that one of the chemistry professors at her school, Dr. Jones, consults for Wow Chemical Company. Susan has applied to Wow Chemical for a job. She does not know Dr. Jones well and has never taken one of his courses. She should visit Dr. Jones and discuss her employment opportunity with him. She should ask Dr. Jones to discuss her abilities and accomplishments with one or two other chemistry professors who know her well. If Dr. Jones is willing, he can then serve as her reference. At the very least, he can comment knowledgeably or provide the names of other professors if colleagues at Wow ask him about Susan's capabilities.

It is often easier for prospective employers to consult "references from left field" about experienced job applicants such as Barry Bench. These may be people the prospective employer knows from trade association and professional society activities. Employed job applicants often do not know these discussions took place. Barry cannot call the "reference from left field" and ask him to regard the conversation as confidential. He can only hope the prospective employer remembered to do so. It is very rare, but sometimes a prospective employer will even contact a colleague at the job hunter's firm to ask for information. Such an unethical act can greatly damage Barry's job security. The negative possibilities of these situations make it critical that applicant inform the prospective employer that her job search is confidential. The job hunter should do this in her initial communication with a prospective employer or headhunter. She should also remind the employer as appropriate during each stage of the applicant screening process.

REFERENCE ETIQUETTE

You should assemble a list of references for each type of job opening you intend to pursue. A phone call is usually preferable to a letter since it allows immediate two-way communication. Explain your situation to your candidate reference. Describe the type of position you are looking for and any accomplishments of which he may be unaware.

Request job hunting and career advice. Ask the person to serve as a reference. If he agrees, send him a copy of your resume. (If he does not agree, but is pleasant, send him a copy of your resume anyway. Ask him to inform you of any relevant job opportunities he hears about. Your resume will help him decide what job opportunities to refer to you.)

As employment opportunities develop with various firms, keep your references informed of your progress. A generic letter you can customize by using a word processor may be the most efficient way of doing this. Telephone your references when you believe a particular employer may contact them soon. When this happens, brief your references on the job requirements. Then they can emphasize the most pertinent aspects of your abilities and experience during discussions with your potential employer.

REWARD TIME

When you get a job offer, discuss it with your references. Listening to their account of conversations with the prospective employer may be helpful in identifying additional questions you need to ask the employer. These could cover issues that need to be settled before you decide to accept or decline the job offer.

Should you accept the job offer, celebrate by taking your references out to lunch. If they live far away, send a card after telephoning them with your news. If you turn the offer down, explain the reason your references. Your references have gone out of their way to help you and deserve to be treated with consideration and respect. Be alert to opportunities to do them favors. Do not forget or neglect them after you start your new job. After all, it is possible you will be job hunting again and will need their help again in the future.

16
Employment Interviewing

Even an outstanding resume will only open the door to possible employment; the interview decides whether or not you get the job. Therefore, the employment interview is the last hurdle in obtaining a job offer. This, of course, is your goal as an employment candidate. Everything you say and do during the interview should help you achieve this goal.

In an employment interview, you are both the sales person and the product being sold. To get a job offer, focus on what you can do for the employer, not what the employer can do for you. Issues such as salary and employee benefits should wait until after receiving the job offer. What to do when the employer raises these issues is discussed below.

The interviewers will be trying to determine the answers to two broad questions during your meeting:

1. Do you have the technical skills to do the job?
2. Will you be a pleasant, helpful coworker who fits in well with the corporate culture?

In all interviews, try to make your answers to specific questions help the interviewer answer these two questions favorably.

Your second goal should be to determine if the job opening is suitable for

---

◆ Some Questions to Ask Yourself

In selling yourself, develop answers to the following three questions:

1. How can I make it easy for the employer to decide to hire me?
2. Which skills, experience, and personality attributes are most relevant to the job opening? And how can I emphasize this?
3. How can I give the impression I can do the job, want the position, and really like the job and the employer?

you. Can you be happy here? Ask the appropriate questions to determine the answer to this question. Job responsibilities and procedures are very suitable items for discussion during your employment interview. Definition of job responsibilities, reporting requirements, and how project teams function are examples of legitimate concerns that you should understand before accepting a job offer. The best and most convenient time to collect information such as this is during an employment interview. However, should you have *serious* concerns, wait until you receive a job offer to raise them. At that time you will be negotiating from a position of greater strength since the company has committed itself to you.

When inviting you to an interview, the employer has decided that your education and experience could make you a valuable employee. His goal during the employment interview is to decide if this evaluation is correct. However, employers will have several other questions of equal importance. Are you willing, cooperative, and enthusiastic? Will you accept and even seek responsibility? Will you get along with your coworkers? What is your growth potential in the job?

Graduating students usually have little interviewing experience. Many mid-career chemists in the job market learn that their interviewing skills are rusty. Even experienced sales personnel often have trouble selling themselves in the environment of an employment interview. There are steps you can take to brush up without having to learn from your mistakes and perhaps losing employment opportunities. Begin by understanding the critical importance of the interview.

PREPARING FOR INTERVIEWS

Before preparing for an interview, job applicants have to understand the interviewing process. The typical chemistry job interview is a series of discussions lasting several hours or more. These discussions include meetings with supervisors, staff members, and a representative from the human resources department. At large technical centers, applicants seeking R&D positions often present a 30- to 45-minute seminar. A carefully organized, well-presented seminar can be the key to getting a job offer.

Managers often will ask applicants technical questions. Beside assessing technical competence, they also will be the using applicant's replies, manners, and body language to assess how well they will fit into the corporate culture. One critical issue is how well the applicant works in project teams. Staff chemists usually focus on the applicant's technical competence. Their concern is whether the applicant will be a competent, cooperative coworker who can meet deadlines.

There is usually a single hiring authority. However, input from all the interviewers is important in making the hiring decision. Applicants must be sure they present a consistent picture to the various people who interview them. Conflicting remarks made to different individuals can raise doubts about an applicant's character, honesty, or motivations.

Target your interview to specific employers. This can be as important as targeting your resume. When writing a resume, and when preparing for an employment interview, it is vital to take the time to know yourself. List your accomplishments, interests, strengths, and weaknesses. Do a reality check by discussing these with someone who knows you well and whose opinion you respect. Review your professional experiences and career goals. This self-assessment will also help you determine what type of job you want. Include your resume in your reading. Be sure anything you say during the interview is consistent with it. Prepare and rehearse discussions illustrating how you solved technical problems, increased sales, and reduced costs. Also be prepared to describe situations where you demonstrated teamwork or leadership skills. Focus on examples likely to be most relevant to this particular employer.

When you have a scheduled interview, review this information. Select items likely to be of specific interest to the employer. Review your correspondence with the employer and articles about them in magazines and newspapers to help you identify these items. On-line databases can help you find such articles. Study the company's latest annual report. Read the Dun & Bradstreet and the Standard & Poor evaluations of the company and its financial outlook. You will learn whether the company is in a growth industry, whether it is well managed, its financial status, and its future outlook.

Become familiar with the areas of technology the employer focuses on. Trade journal articles can tell you what technical fields the employer is emphasizing. Should it turn out there are no opportunities in this field, at least you

◆ Identify Your Interviewers

Before the interview, ask for the names and titles of the people who will be interviewing you. Some companies routinely provide this information in a preinterview package mailed to the applicant. Use managers' titles to anticipate the type of questions they will ask and the information they want from you. Use *Chemical Abstracts* to search for the names of research managers and staff researchers and discover their technical interests. Read their recent publications. Finding a way to smoothly mention these interests during the interview will show that you have done your homework.

will have shown that you have studied the company. Be sure you can discuss your items easily and fluently in an organized way. However, do not memorize a set speech. It takes a lot of practice to recite a memorized speech convincingly. In some cases, the company may be too small for coverage by national publications. Privately held companies do not prepare annual reports. If you have an interview at a small company and were unable to learn much about it, focus on the situation in its industry. If information about a particular company is not available to the public, the interviewer cannot expect you to know about it.

Contact acquaintances who are current employees of the company, particularly people in your field. Diplomatically ask how they like their job and what the advancement prospects are. As you read, prepare questions to ask about the company. These will better define the employment situation for you. Also, good questions show that you have done your homework. Make sure these are relevant, but not critical, of the company. For instance, do not ask why the company's growth rate is less than that of its competitors. Instead, ask what steps the company plans to increase its growth rate.

Preparation for employment interviews is time-consuming. Schedule employment interviews so that you have adequate time to prepare for each. For the employed job hunter, it also means finding the time to prepare carefully. This may require taking vacation time or sacrificing a weekend. Many interviewers are poorly prepared and have received little or no training. The more uncomfortable they are, the poorer their impression of you will be. To have a good interview, you may need to help them. In particular, have a succinct, 60- to 120-second summary of your abilities and accomplishments prepared and find an occasion to use it. Learn about the company and seek opportunities to cite examples of how the company could benefit from your skills.

QUESTIONS

The biggest single determinant of whether you get a job offer is your response to questions. Use the effective listening skills described in chapter 4 to be sure you understand the interviewer's questions. Provide the interviewers with occasional cues such as nodding and affirmations such as, "Yes, I understand what you mean," to indicate that you are alert and paying attention. Except for questions prohibited by federal law and regulations, you should be open and frank in answering questions. Do not lie but be sure your answers present you in the best possible light. Take a few moments to think through your answer before speaking. A focused, organized, pertinent answer has more impact than a rapid one that may be less organized and rambling.

Try to answer questions to illustrate what values are important to you and how you go about solving problems. For example, suppose someone asks you, "Why did you decide to go to the University of West Lumbago for your graduate work?" Do not respond by saying something like "Because my undergraduate research advisor recommended it." This tells the interviewer little about you. Instead, say, "My undergraduate research advisor told me a lot about the school. I learned more about West Lumbago than I could from just reading their literature. I liked the program. When the time came to make a decision, I felt I knew much more about West Lumbago than the other schools I applied to." This kind of response tells the interviewer you like to collect the information you need to make a decision, consider it carefully, and then make an informed choice.

Careful consideration of your circumstances and of the job opening in question will enable you to anticipate many of the questions interviewers will ask. Prepare responses to these questions. Some questions will probe uncomfortable areas such as why you left your last job or got a poor grade in a particular course. For these, prepare responses that are unemotional and factual while presenting you in the best possible light. Also include remarks that will smoothly move the conversation onto another subject. An example would be a short explanation of why you left your last job. Keep this discussion unemotional and focused. Conclude by remarking that you used the time since then to take a short course or write a review paper. Then ask what the hiring company's policies are toward these activities. If possible, note how these activities could benefit the employer. Done well, this will smoothly shift the topic of conversation.

DIFFICULT QUESTIONS

The above example—why you left your last job—can be a "killer" or difficult question. A poorly thought-out answer can destroy your chances of receiving a job offer. You need to review your background to identify possible killer questions and prepare focused, unemotional responses.

If an interviewer spots a discrepancy between information provided in your resume and what you say during the interview, this can result in you facing such a question. The questions asked to resolve this discrepancy can be difficult to answer and stressful. Therefore, your best strategy is to avoid the situation by being thoroughly familiar with all your correspondence and conversations with employer representatives.

Interviewers ask these types questions for two reasons. First, the answers are valuable and revealing. They can significantly help in deciding whether to make you a job offer. Second, the interviewer wants to see how you react under stress. Be ready to answer killer questions that are commonly asked during interviews. These include:

- Why did you leave your last job? Why did you stay only six months at XYX Corp.? Why do you want to leave your current job?

- Why do you want to work here?
- What sort of job do you see yourself doing five years from now?
- Are you willing to travel? How much?
- What aspects of your current or last job do you like best? Which do you like least? Why?
- What is your ideal job?
- Tell me about a coworker you dislike. Why do you dislike this person? Describe your working situation with this individual.
- What do you think of your boss?
- How did you get along with your supervisor? Describe a situation where you had a strong disagreement. How was this situation resolved?
- Describe your role on a project team. What were the roles of the other team members? Discuss a situation where you took a leadership role on the team. Describe a situation where you had a strong disagreement with fellow team members. How was this situation resolved?
- What are your strengths and weaknesses? What don't you do well?
- Why did you become a chemist (or engineer or technician)?

For students, some of the questions vary a bit:

- What was your favorite course? Why?
- Who was your favorite professor? Why?
- Describe a situation, either in class or an extracurricular activity, where you took a leadership role?
- Describe your undergraduate research project. What was your favorite part of the project? What aspect of the project do you like least?
- Describe a typical workday.

Carefully and dispassionately reviewing your situation can identify additional killer questions. Examples include:

- Why do you think you have been unemployed for six months?
- What have you been doing to keep busy while you look for a job? Why aren't you going to night school to improve your credentials?

Rehearse your answers to killer questions likely to be asked during your interview. Tailor your answers to what you know about the job opening.

The seemingly innocuous question "Why don't you tell me about yourself?" can be the most important issue you face during a job interview. This is often the first subject interviewers raise. Your interviewer is looking for a concise, focused review of your accomplishments. You cannot ad-lib an answer to this question, so plan your response in advance. The best way to do this is to list the points you want to make chronologically. Graduating chemists should begin their list with undergraduate school, unless they did something outstanding in high school. Younger, experienced chemists should probably also begin with undergraduate school. Midcareer chemists should probably limit themselves to their work experience. Your response should be about two minutes long. The interviewer can then ask follow-up questions if he or she wishes.

◈ The "Tell Me Your Biggest Weakness" Trap

Interviewers have heard the stock responses so many times, they have become clichés. These include:

- I work too hard.
- I have trouble saying no to customers.
- Once I start a task, I can't finish until it's done.
- I take on too many responsibilities.

It is better to describe a former and real weakness that you have overcome. Describe what you did to eliminate this weakness. This presents you as a disciplined and committed person.

Likewise, your response to the request to "describe a typical workday" reveals a lot about yourself and your current job. It indicates the nature of your interactions with coworkers, the importance of teams in your work, how much time you spend with customers, and your time-management skills. Again, a focused, concise response is best. Avoid the temptation to take a long time to present too much detail.

Responding to "What are your long-term goals?" also reveals much about yourself and indicates whether your career plans and the employer's culture are compatible.

Someone may ask you questions about your salary requirements. If pressed, provide a range, but say that specifics depend on job responsibilities, advancement opportunities, relocation costs, and the area's cost of living. Also, a human resources representative sometimes will ask salary questions to learn what the going rates are for people with your qualifications. Then there are the questions that are more confrontational and stressful. People often ask these to see how you will react under pressure rather than out of any interest in the answer. These sometimes include inquiries on salary requirements. Some of the above questions fall into this category. Too many confrontational questions and you will find you are in a stress interview. Fortunately, stress interviews have become very uncommon. When in one, the best response is to stay focused. Remind yourself of your goal: to receive a job offer, and try not to let the situation upset you. Do not become defensive. Look for a way to move the conversation onto another subject. If more than one interviewer is involved, this says something about the corporate culture. It can be very satisfying to survive the stress interview and receive a job offer. However, in deciding whether to accept this job offer, ask yourself if you can be happy working for the person who ran the stress interview or a firm that would countenance a series of stress interviews.

SITUATIONAL QUESTIONS

Situational questions can also provide information on how you might behave in work situations. In a situational question, the interviewer describes a situation and then asks how you would react. For example, the questioner could briefly describe a project and then ask how you would respond if your supervisor moved up a six-week deadline to only two weeks. The description of how you would reorganize work, obtain the help of coworkers, and ask your supervisor for additional resources could indicate your work methods and how creative or resourceful you are.

Similar questions look to the past rather than the future, but also can reveal a lot about the job candidate. For instance, you could be asked what you think your most important accomplishment was on your last job. Having described something, the interviewer could ask your reasons for picking this project and not other work mentioned in your resume.

It is difficult to anticipate most of these types of questions. Hence, your ability to think quickly and formulate a focused, intelligent response is being evaluated. When asked a difficult question, take a moment to compare your answer. Then, breathe deeply and respond.

THE TAG TEAM INTERVIEW

A new type of interview that is becoming more common is conducted in a way where the candidate meets with more than one interviewer at the same time. The multiperson or "tag team" interview approach can be stressful. The pause by one interviewer to consider her next question may be filled by a second interviewer who asks a question. The conversation is more likely to jump from subject to subject and back again. This makes it harder to prepare considered responses and emphasize points you want to make. It is somewhat akin to the question period that follows an employment interview seminar.

This type of interview has advantages and disadvantages for both the employer and the candidate. Depending on the employment situation, it may be a way for more employees to talk with the candidate during the course of an interview day. It can be less confusing than moving from one office to another for a series of short interviews. It can spare the candidate the strain of a long interview with an unprepared interviewer.

The tag team interview is just becoming popular. Whether it will achieve widespread use or be discarded as not achieving the hiring company's goals remains to be seen.

MOCK INTERVIEWS

Much information is conveyed nonverbally. Therefore, it is critical that your nonverbal communication reinforces the impression you want to make, not detract from it. Mock interviews can help you do this.

Despite all your interview preparation, you may still be nervous, particularly if it has been a long time since you were a candidate in an employment interview. Having an interview go badly can also sap your self-confidence. Mock interviews are an excellent way to gain confidence and improve your interview skills. A skilled interviewer is critical to help you do this. In addition to providing a more realistic interviewing experience, experts can give you excellent feedback and advice on verbal and nonverbal mannerisms, as well as the content of your discussion.

Videotaped mock interviews can be very worthwhile. You will learn about weaknesses you never dreamed you had, such as distracting mannerisms and annoying speech habits. You will also learn how to better employ your interviewing technique strengths. If you videotape a mock interview, be sure your interviewer asks you to "tell me about yourself." Review your response to this critical interview question. Videotaping increases the value of mock interviewing. Rather than just hearing about your technique strengths and weaknesses,

---

◆ Interview Questions that Should Not Be Asked

There are questions that federal equal employment opportunity law and regulations do not permit to be asked. Unfortunately, you may encounter these questions anyway. These relate to how your gender, marital status, family situation, ethnic heritage, and (in some cases) physical limitations and age would affect your job performance if you are hired.

What should you do when you encounter such prohibited questions during an employment interview? Remember your goal is to get a job offer, so you probably should not confront the questioner or report the situation immediately. Instead, say that you cannot imagine how such a situation could impact your job performance. If asked such questions, was it one individual's stupidity or the corporate culture that led to the question? Correctly determining this is critical to your future happiness if you choose to work for this company. It is best to wait until after the interview, review the situation, and determine who is to blame when you are less distracted and emotionally calmer. Remember that some inappropriate questions can arise out of friendly interest, particularly when asked during a meal when the situation is more relaxed.

However, a questioner's ignorance of the law is no excuse. A responsible employer makes sure all its interviewers are aware of what questions cannot be asked during employment interviews. This is true no matter what the interviewers' usual job assignments.

If only one individual asks inappropriate questions, you may decide to accept a job offer from this company anyway. (However, if you do so, you usually need to be sure this individual will not be your supervisor or someone you have to work with closely.) If you believe an inappropriate question is the result of the corporate culture, it is unlikely that you would be happy working for this employer. You probably should inform the company that this behavior has occurred. However, be aware that your statement will probably be an unsubstantiated one. Unless you are deeply insulted and willing to sacrifice any possibility of a job offer from this firm, it is best to wait until your future employment status is resolved. Receiving a job offer and citing inappropriate questioning as a reason for declining it will add force to your complaint. It is best to avoid naming names because of possible libel situations. Of course, if you do not receive a job offer, the employer may perceive your complaint as "sour grapes."

you can actually see them. As an expert shows them to you, it is much easier to identify ways to improve your techniques. You can make a series of tapes as you work on improving your interviewing techniques. This can help you see your progress in improving your skills. Even if there is still a need for improvement, this progress can increase your self-confidence.

THE INTERVIEW SEMINAR

Academic and industrial employers often ask both advanced degree graduates and experienced scientists to present an employment interview seminar. Its purpose is to help the employer determine if the candidate can be a valuable member of the R&D team. The interview seminar enables the employer to assess the candidate's technical and communication skills. A candidate's presentation seminar also can reveal important personality traits such as creativity, perseverance, and patience with questions. The candidate's conduct during the interview seminar, particularly the question period, can indicate how they will perform under pressure. The oral presentation skills discussed in chapter 3 can help job candidates organize and present their seminar in a logical, clear, and professional manner. In particular, be sure your host understands what audiovisual aids you will need.

Choosing the seminar subject is important to a successful seminar. If at all

◇ The Impact of Body Language

Your body language during interviews sends important messages. These messages should be consistent with the image you want to convey. Videotaping can help you improve your body language. To send the message that you are a forceful, self-confident person, maintain good eye contact with interviewers. If you let your eyes wander, the interviewer receives a message that you lack interest or energy. Do not make eye contact and then immediately let your eyes dart away. This creates an impression of nervousness.

Slouching suggests you are nervous and perhaps lack self-confidence. Sit or stand erect. (Sitting on the edge of the chair helps.) If you stand while talking to someone, do not lean against anything. Avoid perching on a table or the back of a chair.

Review videotaped mock interviews to help you identify and eliminate any annoying mannerisms. These include tapping your fingers, wiping your hands on your clothing, doodling, playing with a pencil or other object, and many other repetitive activities. These mannerisms suggest you are not really paying attention. Never fold your arms. This suggests a rejection of the ideas or information the interviewer is communicating. Smile frequently. Avoid forced smiles that quickly vanish. This betrays insincerity.

Coordinate your body language. For instance, leaning forward implies interest. But do not then glance around as if you are bored. Shaking hands is about the only mannerism that videotaping mock interviews will not help you improve. Shake hands firmly without the other person feeling his hand is in a vise. If your hands tend to perspire, keep a paper towel or tissue in your pockets. Then putting your hands in your pockets will let you unobtrusively dry them.

Videotaping interviews was once the preserve of expensive placement firms. However, relatively inexpensive video cameras and VCR's have made this valuable technique widely available. The American Chemical Society organizes mock interviewing sessions at national and regional meetings complete with videotaping. Experts provide immediate feedback on how individuals can improve their interviewing skills. Other professional societies offer similar services, as do many university placement offices and some outplacement services.

What if you do not have access to these services or you do not own a video camera. Consider joining a job-hunting club. One of the members is likely to have a camera. You may be able to work through a local section of your professional society. One member can interview you while others act as a silent audience. Chemists and engineers may lack the interviewing and interview analysis skills of experienced employment consultants. However, by having several people review your videotape together, you can receive valuable ideas and suggestions.

possible, candidates should talk about their own research even if it has been published or patented. Issues of confidentiality can make this difficult for experienced industrial chemists. However, a trade association paper, a patent, or other open literature information may provide suitable seminar material. If you cannot present a paper on your own research results, consider presenting a review of technology in your current research field. Of course, graduating chemists can talk about their thesis research.

Unlike a presentation at a technical conference or a team meeting, the primary purpose of the interview seminar is not to educate the audience about a subject. It is to convince them the candidate will be a valuable member of the employer's R&D team. The presentation should demonstrate the candidate has the following qualities:

- basic knowledge in a technical field of some interest to the employer. The candidate should strive to make any relevance to the employer's R&D interests clear. (This is another instance in which reading about the company in advance can improve a candidate's employment chances.)

The seminar should demonstrate that the candidate has:

- judgment in defining a significant research problem that can be solved with reasonable effort. The choice of techniques used to solve the problem is also important.
- energy and determination in performing the needed experiments.
- an ability to change theories when the data so indicates.
- clarity of analysis in interpreting the data to solve the problem.
- clarity of expression in presenting the results and interpretation during the seminar.
- the ability to accept constructive criticism during the seminar question period. The candidate should also demonstrate patience in answering naive or obvious questions while exhibiting maturity in answering hostile questions. If you do not know the answer to a question, do not try to bluff your way through. Admit you do not know and promise the questioner you will find the answer. Get the questioner's name and telephone number. Get your questioner the answer as fast as you can.
- honesty in giving appropriate credit to mentors and coworkers.
- enthusiasm for science or engineering and enjoyment of the project.

Candidates should choose the topic and organize the seminar to emphasize these qualities. While it is best to strive for relevance to the employer's interests, graduating chemists often have few options in this area. In this situation, do not become obsessive about relevance; the employer should understand the constraints you face. However, do look for relevance as the employer invited you to interview for a reason.

Experienced candidates should try to describe an entire R&D program from conception to commercialization while emphasizing their role on the team. This

◆ Ten Major Interview Mistakes

Short of showing up on the wrong day or spilling coffee on the hiring manager, the biggest interview mistakes are things you do that you can usually spot during a mock interview. However, you can do this only with a person who is both a good interviewer and a good role player conducting the interview. These mistakes are:

1. Not presenting a confident (but not arrogant) image. Letting your nervousness show by speaking in a monotone, avoiding eye contact, and having sweaty palms can kill your chances of employment.
2. Not answering questions to help the interviewer understand your thought processes and how you make decisions.
3. Forgetting that it is important to demonstrate your ability to fit in with the corporate culture. Being observant during interviews and tours and networking with employees or people in the same industry can help you understand the basics of the employer's corporate culture.
4. Not determining what qualities and abilities the company wants in a new employee and then demonstrating you have them.
5. Not focusing on values, skills, and accomplishments. During the interview be sure you cite specific examples of your abilities and accomplishments.
6. Responding to questions in a rambling way—particularly the request, "Tell me about yourself." Be focused and concise (but not curt) in answering questions.
7. You did not study the company and know its major businesses. This often becomes apparent in interviews. Once it does, this spells doom for the applicant. Because you do not understand the company, you cannot cite good reasons why they should hire you.
8. You did not come prepared to deal with questions about your lack of experience. Both students and career changers are particularly susceptible to this issue. Come prepared to state why your abilities compensate for a lack of experience.
9. You did not prepare for a weak interviewer. If it becomes apparent that you have one, near the end of the interview, be forceful and state why you think you are an excellent choice for the job.
10. You forget you are interviewing for a job. Instead, you relax and just have a pleasant conversation. This can easily happen with poorly prepared interviewers.

will demonstrate a variety of valuable abilities to the employer—both technical and organizational skills.

An alternative for the experienced chemist is to present a critical technical review of a particular subject of interest to the employer. Candidates should use this option only as a last resort. Unless you offer extensive new interpretations, the opportunity to demonstrate your own creativity may be limited. By choosing a subject relevant to the employer, you may find members of the audience have expertise in the subject material. They may take issue with your interpretations or conclusions.

In responding to questions, you should be open and friendly. However, do not allow a technical disagreement to develop into a long discussion with one member of the audience. Should this begin to occur, suggest that you can pursue the subject later in the day or discuss it by telephone after the interview. Before opening the floor to questions, experienced candidates should take the opportunity to summarize their career. This can include courses of study, to unique industrial experiences, such as acting as a liaison with patent attorneys for your department, and so on. Candidates active in professional societies might mention this and briefly note how these activities benefit their current or previous employer.

An excellent interview seminar can be a winning situation for candidates even if they do not receive job offers. The employer may decide the candidate would make an excellent employee, but in a different position. Should an appropriate position open, the employer may contact the candidate later. Some members of the seminar audience may form an excellent opinion of the candidate. They may inform you of employment opportunities elsewhere in their company or at other firms. Then network with the candidate later.

Videotaping and reviewing your interview seminar can help improve your public speaking skills (see chapter 3). Graduating scientists and engineers can also improve their presentation skills by presenting their interview seminar to their professor, research group, and friends. The questions they ask may anticipate those asked during the on-site interview.

FINAL INTERVIEW PREPARATIONS

The old saying, "you never get a second chance to make a first impression" is certainly true for employment interviews. Proper attire is critical to employment interview success. Strive for a neat, tailored look. Applicants should also not wear large, bulky jewelry.

Both men and women should wear a dark suit: gray, navy blue, or perhaps a black pinstripe. Men's suits should be single breasted. A white shirt or blouse is best. Women should avoid bright colors except an accent such as a neck scarf or handkerchief. Women should wear black or dark brown shoes with a short heel. Belt and purse should complement the shoes. Men should wear a silk tie and avoid garish colors and patterns. A breast pocket handkerchief is optional. It should complement, but not match your tie. Wear well-shined lace-up black or dark brown leather shoes and black or navy blue over-the-calf socks. Belts should complement the shoes.

If you travel out of town for the interview, pack your clothes carefully so they do not wrinkle. Should you fly, use a carry-on bag for your clothes. Carry a handkerchief in your pocket in case you develop sweaty palms. Bring breath mints and use one before your first interview.

Bring extra copies of your resume with you to the interview. If you have written articles that were recently published in trade magazines or technical journals, bring copies to the interview to give to interested parties. These can help define common interests with current employees. Have names, addresses, and telephone numbers of your references available. You may be asked to complete an employment application, so also bring the dates of employment, addresses, and phone numbers of previous employers.

THE INTERVIEW DAY

Arrive at the interview on time. No excuse, no matter how valid, compensates for the bad impression a late arrival makes. Allow time for transportation delays, particularly in winter. Should a late arrival be unavoidable, telephone as much in advance as possible to inform the employer of the delay. If you are driving to the interview site, get directions from your host. If you are in an unfamiliar city, check the typical traffic situation on your route with the hotel doorman or rental car agent.

Sometimes a host will begin the interview by taking you out to breakfast. Should this person meet you at the hotel, be ready in the lobby; do not force your host to telephone your room. Should you meet your host at an airport or busy hotel lobby, alert your host in advance of your appearance. An alternative is to inform your host that you will be carrying a distinctive item. Making yourself easier to recognize simplifies your host's job.

◆ A Job Offer Can Increase Future Job-Hunting Confidence

Even if the interview begins to indicate this is not a place where you would want to work, do not relax. Continue to go for the job offer. The emotional satisfaction a job offer entails can re-energize your job-hunting efforts by improving your morale. It can improve your self-confidence and performance in subsequent job interviews. In a large company, the unsuitable position in one department does not necessarily mean that different positions in other departments are unsuitable. An excellent interview day for the first position can make you a leading candidate for other positions that may be more suitable.

Remember people's names. Repeating names when you are introduced and using them in later discussions will help. Some companies provide applicants with an interview schedule before the interview. Review and memorize the names and job titles of the people you will meet. Ask the names of receptionists or secretaries you meet. Be polite to them when you arrive on site. Wish them a nice evening when you leave. If the interviewer observes, so much the better. If not, a later "what a nice person" comment from the secretary or receptionist to the interviewer will not hurt.

Be lively, enthusiastic, and cheerful. Remember to smile. People want energetic coworkers with pleasant dispositions. Demonstrate that you are one of these people. An all-day or half-day series of discussions can be quite tiring. Therefore be sure you get a good night's sleep before the interview. If you have to fly to another city for the interview, it is usually best to arrive the night before and stay in a hotel. (This also makes it easier to be prompt.) However, if you have difficulty sleeping in a strange room, it may be preferable to take an early morning flight.

Do not smoke, even if the interviewer lights up. (A survey by the executive recruitment firm, Robert Half International, finds one in four smokers will be rejected solely for smoking when competing for a job with an equally qualified nonsmoker.)

Request permission before moving a poorly positioned chair. If left alone in an office, do not wander about. If observed, people may think you are "nosy." Remain seated and consider how the interview has gone so far. Review points you want to make during the discussions. If the interviewer has left company literature with you, skim through it. If there are plaques or certificates on the wall, read them. They may tell you about the interviewer's education, awards, and professional activities.

Remember, your primary goal is to get a job offer. Every question you answer or ask during a job interview should move you toward this goal.

Can you better determine what aspects of your experience and education are most relevant to the job opening? Remember the request "Tell me about yourself" (see above)? When you complete your answer, ask the interviewer, "Which area would you like to discuss first?" His answer will tell you what is most important to the job opening. During discussions, bring up aspects of your education and experience that are most relevant to the job opening. Do this at every opportunity. Interviewers, particularly managers and those from human resources departments, want to know about your interpersonal skills. Of special interest is the ability to work well as part of a project team. Self-motivation and initiative are also important qualities.

Avoid criticizing former employers, supervisors, or coworkers. Focus on the benefits you provided to these employers. Discuss situations that emphasize your interpersonal skills. The description of technical skills in your resume helped persuade the employer to interview you, however, it is much harder to assess interpersonal skills. The most important of these are teamwork and leadership skills.

You should have already prepared questions about the employer. During the discussions, ask these questions. They demonstrate your interest in the company and show you had the initiative to learn about it. If you interview with more than one person, do not ask each the same list of questions. However, ask more than one person your most critical questions: those having to do with job satisfaction, job security, and career advancement. Listen carefully to the interviewers. Before answering a question, be sure you understand it. If you do not, ask for clarification before answering. Also, the wording and tone of questions may give you clues about the answers interviewers expect or want.

Do not lie or compromise your principles, but do try to respond so each answer indicates that you will make a valuable employee for the company. Avoid a direct answer to questions about your salary requirements. Your goal is to get a job offer; you can negotiate your salary later. Hiring authorities will be more willing to meet your salary request if they have already made you a job offer than if they have not. If pressed, say that you expect a competitive salary. As suggested above, provide a salary range. Comment that your exact requirements will depend on the responsibilities and advancement opportunities of the job. Explain that you are still learning about these and that your interview has given you a lot to consider. Therefore, you cannot name a particular salary figure. Then switch the subject back to the job requirements. (A human resources

---

◆ Interview Meals

On-site interviews often last several hours, so there is a good chance that you will be invited to at least one meal. If so, relax, but remember you are still being evaluated. Meals provide a good opportunity to assess interpersonal skills and compatibility with the corporate culture.

Use your best table manners. Avoid dishes that are difficult to eat or are dripping with sauces that could stain your clothes. Avoid dishes with stringy cheese such as French onion soup and some Italian dishes. Eat lightly. A large breakfast or lunch could leave you sleepy later in the day. Do not order an alcoholic drink, even beer, during lunch—even if your host does.

After the meal, praise the restaurant and thank your host. Even if the food or the service was poor, avoid criticizing. Should your host criticize the meal, voice mild agreement and change the subject.

representative may ask you what salary offers you have received from other companies. They are doing this to understand how the job market values your qualifications. When you respond, again give a range and do not identify the companies that have made you job offers.)

If you are thirsty during interviews, accept offers of coffee or soda. However, beware of drinking too much coffee. The caffeine lift could vanish quickly and leave you feeling tired.

Careful preparation will improve your self-confidence during the interview. A confident, but not arrogant, manner will impress interviewers. It will help you respond effectively to unexpected questions. By planning your interview points in advance, you will be able to steer the conversation to your strengths and minimize discussion of your weaknesses.

After each discussion, do more than just shake hands and thank the interviewer. Conclude by restating your strong interest in the job. Say that you believe you are well qualified and can handle the job. Finally, state that you would enjoy working with the interviewer. When you talk to the hiring authority and the human resources representative, go one step further—ask for the job.

SCREENING INTERVIEWS

Screening interviews often precede the on-site interview. The main types of screening interviews are telephone interviews, employment clearing house interviews, and, for students, on-campus interviews.

The telephone interview can come unexpectedly, so you should have prepared for employment interviews before receiving any invitations for interviews. If you have followed the guidelines given above for on-site interviews, it may be best to begin your telephone interview immediately. However, it is acceptable to ask for an appointment for the interview delaying it for a day or so. While this will also give you time to research the employer, your caller probably does not expect you to be highly knowledgeable about the company.

EMPLOYMENT CLEARING HOUSE INTERVIEWS

Clearing house interviews also offer participants little time to prepare. The American Chemical Society, American Institute of Chemical Engineers, and other professional societies organize employment clearing houses for their members. For example, the ACS National Employment Clearing House (NECH) is held in conjunction with both national and regional ACS meetings. These 15–30 minute "mini-interviews" may be familiar to students who participate in on-campus interviews (see below). However, experienced chemists new to the job market can be disconcerted at their first clearing house interviews; I know I was.

Why is the mini-interview different from on-site interviewing? Employment clearing houses provide opportunities for employers to interview many chemists in a short time. The atmosphere is more impersonal as the interviewer has little time to put candidates at their ease or gradually lead into important points. Employers invite only the most suitable clearing house candidates to on-site interviews. The interviewer's job is to identify these candidates. Detailed scrutiny of candidates' accomplishments and abilities is left for the on-site interview. The interviewer specifically has to determine if the candidate's experience and abilities are a good match for a job opening. This can lead to more rapid and intense questioning than may occur in more leisurely on-site interviewing.

The interviewer also needs to look for any obvious personality factors that may make a candidate a poor match for the job opening or the corporate culture. An example is an expressed disdain for organizational procedures. This candidate might do very well at a small company, but have problems in a large, more bureaucratic organization. Open-ended questions that encourage candidates to express themselves are the best way of assessing the candidate's personality. However, the interviewer has little time to do this. For example, standard interview appointments at the ACS National Employment Clearing House are only 15 minutes. However, some interviewers (including this author) schedule 30-minute interviews.

The interviewer will be talking with several job hunters during the day. This makes it is difficult for a candidate to make a strong, favorable impression. The intensity of the interview means the prepared candidate can have a great advantage. But how can you prepare with only a few hours advance notice of your in-

---

◇ Following Up on Clearing House Interviews

An interviewer's invitation to dinner or cocktails for further discussions after an employment clearing house interview during the professional conference is an encouraging sign. If the job opening is of special interest, suggest such a discussion yourself. This demonstrates your strong interest in the employment opportunity.

As these discussions can be very rewarding, be willing to adjust your conference schedule to accommodate them. For example, I received such an invitation and was thus able to spend more time discussing my qualifications. Also, my interviewer gave me a very clear picture of what being an R&D chemist for his employer was like. I received an on-site interview, accepted a job offer, and spent eight rewarding years at Halliburton Services.

As a result of this experience, I often invite promising candidates I interview at clearing house interviews to an additional private interview during a professional society meeting.

terview? You should register in advance or arrive early to have time to complete clearing house forms that detail your professional experience and qualifications. Bring a supply of resumes to provide to interviewers. Do not passively wait to be interviewed. Review job openings listed by companies participating in the clearing house. Request interviews when you believe your qualifications match a listed job opening. Whether you or the recruiter requests an interview, accepting interview requests is voluntary.

In on-site interviews, chemists can learn about the employer before the interview and understand the employer's needs and goals. Such is often not the case at employment clearing houses. The job hunter learns what companies are interviewing only after registering at the clearing house. Therefore, candidates should read about the current business and technical situations in various industries of interest rather than narrowing their focus to particular companies. Of course, if a job candidate who attends an employment clearing house has a notebook computer or other access to the Internet, he could do a quick search and learn something about a company that will be interviewing him.

After scheduling an interview, read the clearing house information on job openings offered by the interviewer. If the interviewer provides corporate literature in advance of your interview appointment, read this also. Check to see if the interviewer or a member of the same firm is presenting a paper at the conference. Job candidates should attend these papers or read the abstracts. These will define an area of technology of interest to the employer. An alert candidate may also pick up clues about corporate culture from the speakers' remarks. Pertinent remarks to the interviewer about the presentation can indicate your interest in the employer, as well as convey a sense of professionalism.

The interview invitation will specify an interview table number and time. Be sure you are on time. Locate the table in advance so you do not waste interview time searching for it. A brisk direct walk to the interview table plus a firm handshake suggests self-confidence. Dress as you would for an on-site interview (see above). Focus on the interviewer and ignore the buzz of conversation from nearby interviews. This is easier said than done. During the limited time available, candidates should emphasize their technical accomplishments of interest to the employer. Carefully review the firm's listed job openings to help identify what to emphasize. End the interview by expressing a strong interest in the job opening. Ask if you can provide any additional information. Determine when you will be hearing from the recruiter. If the interviewer is not sure, suggest that you call the interviewer after two to four weeks have passed.

CAMPUS INTERVIEWS

On-campus interviews share features with both the on-site and clearing house interviews. Like on-site interviews, on-campus interviews are scheduled well in

◆ Video Interviews

One reason a firm does not recruit at a particular campus is there are only a small number of students with the needed skills at this college. Colleges and employers are coping with the cost of this problem by using video conference centers to conduct interviews.

Both interviewer and candidate can see each other and pass documents such as resumes and application forms back and forth via fax machine. Video conferencing can be disconcerting. For example, with current technology, there is a two-second delay between when a candidate sees the interviewer begin to speak and when he hears her. Improving technology should reduce this disconcerting delay.

By mid-1995, more than 70 campuses had installed video interview equipment and software. Cost of the equipment and software is approximately $5,000. Video phone bills average about $60 per hour. Companies without a video conference center of their own can rent time in commercial facilities. While less personal than a face-to-face interview, video interviews are cheaper than campus visits. Colleges are finding that smaller employers, often companies that never recruited on campus before, are making use of video interviewing.

advance so that job candidates can read about the employer and prepare relevant questions. Like clearing house interviews, on-campus interviews offer employers the ability to see many candidates in a short time and screen then for on-site interviews. Also like clearing house interviews, on-campus interviews are short—usually 30 minutes.

◆ Screening Interviews

Telephone, video, clearing house, and on-campus interviews are all screening interviews. At this stage of the process, the employers are often more interested in negative information that will aid in eliminating candidates, rather than in positive information that will help them make hiring decisions. This is reflected in the questions they ask.

At this stage of the employment process, the hiring authority is like a coach cutting unqualified players from a sports team. Based only from the on-site interview will they choose employees from the candidates who remain.

Interviewers want to judge whether you will be compatible with their corporate culture, so be prepared to talk about more than the technical subject. Ask your placement office about the background of your interviewer. The interviewers' background can determine both the questions you ask and how you respond to her questions. For instance, suppose someone asks you to describe your senior research project. If your interviewer is a research chemist, you should provide more technical details than if you are talking to someone from the human resources department.

JOB FAIRS

In large cities, local employers often participate in employment fairs. Local newspapers often sponsor these events. They are held in local convention centers and often draw large crowds. These are often unfocused, with companies looking for candidates for many types of positions. Chemists, engineers, and technicians find that other job hunters are accountants, computer specialists, and members of other professions.

The employers usually stand at booths where they answer questions and provide company information. Interested candidates can complete employment forms and leave resumes. There is usually little time for any real interviewing. Conversations can last several minutes at most. However, employment candidates should dress as if they were going on a conventional interview and conduct themselves in a business-like manner. If discussions at the booth indicate a job opening interests you, ask the company representative for an on-site interview appointment.

Job fairs are often useful opportunities for chemists to make contacts and explore alternative careers. For instance, you are a synthetic organic chemist at a job fair. You see an engineering firm is looking for technical writers. This is a good opportunity to find out the responsibilities of a technical writing position, the salary, and other pertinent information.

INTERVIEWING FOR FACULTY POSITIONS

Much of what has been said above applies to interviewing for academic jobs. For example, teamwork skills may be less critical than in an industrial situation. However, they are still important in taking advantage of opportunities to work productively on research problems with other faculty members. University committee work calls for teamwork and leadership skills. Communication skills may be called teaching ability, but are still very important.

The academic interviewing process consists of the presentation of a seminar and a series of meetings with faculty members, the department chairperson, and one or more deans. At some universities, students play a significant role in the interviewing process. A time and room may be set aside for students to interview candidates. Students are often blunter and less diplomatic in asking questions. In front of a room full of people, faculty candidates may find themselves responding to questions such as:

- Why do you want to come here?
- Why do you want to leave your current job?
- What makes you think you would be a good teacher here?

They may ask specific questions about your expectations for student performance. Students are rightly concerned about candidates' teaching abilities, as are department chairpersons. Some faculty job candidates report being handed a course catalog and asked which of the department's courses they could teach and why. This is an example of the importance of preparing for academic job interviews. Advance knowledge of the department's course of study and research interests of faculty members can be very helpful in discussing how you would benefit the university as a faculty member. Knowing the capabilities of interdisciplinary institutes on campus can also be useful, as can knowing what instrumentation is available in other departments that could assist in your research.

Candidates need to demonstrate the ability to define challenging research problems. Demonstrating at least an appreciation of the difficulties in obtaining research funding and the intention to be aggressive in applying for funding is important, particularly when interviewing with department chairpersons and deans.

Midcareer industrial chemists will find themselves at a disadvantage in competing for academic positions. Their inexperience in teaching may cause concern. Many industrial chemists will find their short-term applications research will not impress academic interviewers. In addition, industrial chemists do not have a track record in winning the grants necessary to establish a solid academic research program. (However, they can emphasize their role in winning corporate funding of their R&D program.) For these reasons, it is very difficult for all but the most distinguished industrial scientists to obtain a senior faculty position. At the junior faculty level, midcareer chemists are competing with recent graduates. The perception is that younger faculty members will be satisfied with lower salaries.

INFORMATION INTERVIEWS

These are definitely a different type of interview. *You* will be doing the interviewing. You will be interviewing experienced professionals to find out about employment opportunities in different fields and industries and what their work life is like. Both students and experienced professionals interested in changing fields or exploring employment opportunities in different industries will benefit from information interviewing. The people you interview may become members of your network alerting you to employment opportunities in their fields and industries.

Begin your information interview by selecting fields or industries that interest you. Read trade journals covering industries in which you have an interest. Call or write professional societies covering fields that interest you. Many organizations will have brochures and other information they can send you. This information will provide you with an information base on which to build when you conduct your information interviews.

Then, contact members of your network to collect names and telephone numbers of professionals in fields that interest you. Professional society membership directories can also be a source of names and telephone numbers, as can local sections of professional societies. Call these individuals and request an interview either by telephone or in person. Plan on about 30 minutes for a telephone interview and 60 minutes for a personal interview. Personal interviews are often conducted over lunch. The professional you interview may offer to pick up the lunch bill, particularly for students. However, plan on picking up the bill yourself, even if you are unemployed. After all, they are doing a favor for you.

It may be difficult to find someone who has the time to sit down with you. Most professionals do not have the time for information interviews with job candidates more than very occasionally. If you have problems locating someone to meet with, try working through the local section of your professional organization to identify people who may have the expertise you need and may be willing to meet with you. Be creative. While the stereotypical information interview may occur at lunch, there are other scheduling options that may be more convenient for the person you want to meet with. Breakfast at a convenient restaurant is one example or a meeting immediately prior to a local professional meeting you are both attending.

During your interview, ask about the individual's daily work life: what their primary activities and duties are. Ask them what they enjoy most and least about their jobs. Discuss employment and career advancement opportunities. Do not ask your interviewees about their salaries. You can get this information from professional society salary surveys or the United States Department of Labor Bureau of Statistics. (A helpful librarian can help you find this information.) You may also want to inquire about job-hunting techniques, suggestions on resume preparation, and other job hunting topics. If you are changing careers, show them your resume and ask what aspects of your experience are most relevant to the new career field. You may ask if this person knows of any companies who may be hiring. However, do not ask for a job. This person came to do you a favor and expected to feel no pressure.

Try to conduct information interviews with at least two people from a field or industry. This will help ensure you get a balanced perspective. However, do not become overly fond of information interviewing. During the loneliness of the job-hunt, it can be tempting to reach out for the companionship you can find in information interviews. However, if they seem interested, do occasionally call and keep your interviewees informed of your job-hunt progress. If you need companionship and support, join a job-hunting group.

FINAL REMARKS

Believe in yourself. You will never persuade the interviewer that you will make an excellent employee if you do not believe this yourself. A successful employment interview will leave you with an enviable problem: assessing a job offer and deciding whether to accept it.

ADDITIONAL RESOURCES

Books

The Interview Handbook. American Chemical Society: Washington, DC, 1995.

Allen, J. G. *The Complete Q&A Job Interview Book*. Wiley: New York, 1997.

Corcodilos, N. A. *Ask the Headhunter: Reinventing the Interview to Win the Job*. Plume: New York, 1997.

DeLuca, M. J. *Best Answers to the 201 Most Frequently Asked Interview Questions*. McGraw-Hill: New York, 1996.

Drake, J. R. *The Perfect Interview*, 2nd ed. AMACOM: New York, 1996.

Fry, R. W. *101 Great Answers to the Toughest Interview Questions* 3rd Ed. Career: Franklin Lakes, NJ, 1995.

Kay, A. G. *Interview Strategies That Will Get You the Job You Want*. Betterway: Cincinnati, OH, 1996.

Krannich, C. and Krannich, R. L. *Interview for Success: A Practical Guide to Increasing Job Interviews, Offers, and Salaries*. 7th ed. Impact: San Luis Obispo, CA, 1998.

Medley, H. A. *Sweaty Palms: The Neglected Art of Being Interviewed*. Ten Speed: Berkeley, CA, 1993.

Yate, M. *Knock 'Em Dead: The Ultimate Job Seeker's Handbook*. Bob Adams: Holbrook, MA, 1994.

Videos

"Interviewing." American Chemical Society, Department of Career Services: Washington, DC.

17

ASSESSING JOB OFFERS AND NEGOTIATING BETTER TERMS

Deciding whether to accept a job offer can affect your career for many years to come. This is true whether you have two or more job offers to select from or you are deciding to accept a job offer or stay with your current employer. Of course, it is also true if you are unemployed and have received only one job offer.

Remember, no job, and no job offer, is ideal. In assessing your job offers, consider these factors:

- What is the starting salary? Your starting salary sets the baseline for all future salary increases. Raises are set by company policy; exceptions are few. Therefore be wary of promises to compensate for a low starting salary by fat annual raises. Your starting salary is also the key determinant of the present and future value of your fringe benefits, particularly your retirement package.
- What are your job responsibilities? Are they consistent with your skills, training, and experience? Are they consistent with what the employer told you earlier?
- What are the job advancement and career development opportunities? This is a subject you should have explored during your employment interview, however, if you did not do so before, do it after you receive the job offer. Do not rely on input from just one person. Discuss this issue with the hiring authority, the human resources person you are working with, and any contacts you have inside the company. An excellent way to explore this is to discuss career paths with people who started in comparable positions with the firm.
- What are the benefits? These are listed in the box below. Which are appropriate for you depends on the job for which you are being hired. For example, sales commissions are an important consideration for candidates for chemical sales jobs, but not for research job candidates. Some benefits such as country club memberships are reserved for high-level executives. Which benefits are most important depends on your personal situation. For example, retirement programs will be much more important to a 45-year-old applicant than to her 25-year-old counterpart. The opposite may be true for maternity and family-leave benefits.
- What are the relocation requirements? Which relocation expenses are funded by the employer? Are these negotiable? It can be particularly worthwhile to negotiate unusual relocation expenses if the employer re-

jects another one of your requests, such as a higher starting salary.

- Is the corporate culture compatible with your personality and work style? Are your future coworkers friendly, compatible, and helpful? The best time to assess these factors is during your employment interview. Asking contacts in the company about specific individuals (perhaps your potential supervisor) can be helpful. However, remember that your contacts have divided loyalties to you as a colleague and friend and to their employer who pays their salaries.

Consider the process of comparing two job offers or a job offer with your current job. Begin by listing the important factors of each. You can work from the list in the sidebar. Identify each category where one job offer is superior to the another. Then calculate the total number of categories in which each job offer is superior to another.

When comparing a job offer to your current job, include any financial losses associated with changing jobs. This may include some relocation costs and a

---

◆ Job Offer Factors

JOB RESPONSIBILITIES
   Your compatibility with:
      your manager
      coworkers
   Firm's reputation in its industry

DIRECT COMPENSATION
   Base Salary
   Sign-on Bonus
   Commissions
   Corporate Profit Sharing
   Personal-Performance Bonus
   Compensatory Pay during Military Duty

HEALTH BENEFITS
   Medical Insurance
   Dental Insurance
   Life Insurance
   Disability Insurance
   Periodic Medical Exams

RETIREMENT RELATED
   Defined Benefit Plan
   401(k) Plan

NON-NEGOTIABLE
  Deferred Compensation
  Savings Plans
  Stock Purchase Plans

OTHER
  Stock Bonus
  Stock Options

PROFESSIONAL, EDUCATIONAL
  Continuing Education
  Professional Society and Trade Association Membership Fees
  Tuition Reimbursement for Yourself and Family

RELOCATION EXPENSES
  Direct Moving Costs
  Moving Costs for Unusual Property
  Trips to Find Housing
  Loss on Sale of Present Home or Lease Termination
  Company Handles Sale of Present Home
  Interest-free Loan while Owning Two Homes
  Mortgage Rate Differential between Old and New Homes
  Mortgage Fees, Closing Costs
  Temporary Dual Residences
  Trips to Old Home during Dual-Residence Period
  Spouse Outplacement Assistance

VACATION TIME

PERQUISTITES ("PERKS")
  Company Car
  First Class Travel, Hotels
  Personal Use of Frequent Flier Mileage
  Paid Travel for Spouse
  Social Club Memberships
  Executive Office
  Private Secretary
  Employee Discounts on Company Products

waiting period before you are eligible for some benefits in your new job such as savings plans and stock purchase options. Of course, comparing job offers is more than just comparing a score. Some factors are more important than others. Nevertheless, going through this analysis will enable you to focus on the critical comparison factors. For example, only you can decide if the greater challenges offered by Job Offer 1 outweigh the higher salary of Job Offer 2.

NEGOTIATING YOUR JOB OFFER

Before weighing the positive and negative aspects of the job offer and deciding whether or not to take the job, negotiate to increase the positive and decrease the negative. Do not assume any parts of the employment offer are "set in stone." All may be negotiable. However, only negotiate those that are the most important to you, such as salary or relocation benefits.

Your negotiating strength is never greater than when the employer has made their decision and offered you a job. Now the employer is waiting for you to make a decision. Of course, they want your decision to confirm and validate theirs—that is they want you to accept their job offer. Until you do, you are an individual—a person in whom the company has made a financial and emotional investment. Once you accept, future raises, promotions, and so on must conform to standard corporate guidelines. You are then part of a group and your situation is no longer unique.

This means it is critical to negotiate improvements in aspects of the job offer with which you are dissatisfied before accepting the job. If you wait until after you accept the job offer, you will have waited too long and the employer will probably be less flexible. You should handle these negotiations in a professional, nonemotional manner basing your negotiating position on facts. Even if they cannot accommodate you, negotiating in a reasonable, responsible manner can raise their opinion of you.

Negotiation abilities are important in many job assignments, so you have just demonstrated you have another useful job skill. However, you do not want to seem argumentative and difficult to get along with. Do not negotiate just for the sake of negotiating. Concern yourself only with important issues.

Before beginning your negotiation, be certain that you understand all aspects of the job offer. Without this understanding, you may ask for something that is already part of the offer. Clarify any points that are vague before beginning your negotiations.

Job Responsibilities

These are usually not negotiable. However, be sure the job responsibilities are the same as detailed in your employment interview. If you feel some added responsibilities are inappropriate, first consider whether they may be opportunities for personal growth. If they are, accept them, but ask the employer to commit to a training program to enable you to better fulfill these responsibilities. Less commonly, the job is narrower in scope than first stated. Again, determine the reasons for this and be sure they are acceptable to you.

Suppose some responsibilities were not part of the package when you interviewed and you would dislike the work necessary to fulfill these responsibilities. In this case, negotiate to have these responsibilities omitted from your job assignment. If they cannot be omitted, politely determine why they were added to the job description. The answer can tell you something about the character of the supervisor and perhaps about the corporate culture. If the reasons employers give for any changes in the job responsibilities are not good ones, this can indicate possible future problems with arbitrary management decisions after you accept the job offer. Of course, if the added job responsibilities are things you strongly dislike, you may wish to decline the job offer.

Salary

Virtually all employers have a salary range in mind when they offer you a position. Except for entry-level positions, they are not likely to make their best offer at first. Judge the salary offer in terms of your skills, experience, and personal situation. Obtain advice from trusted peers and mentors. Determine the cost of living in the job site area. The annual ACS Salary Survey published in *Chemical & Engineering News* is also helpful. Some professional engineering societies also perform and publish salary statistics as does the National Science Foundation and the Bureau of Labor Statistics.

Suppose the salary offer is less than the going rate for your skills and experience. There may be a good reason for this in other aspects of the employment package such as an up-front sign-on bonus or a bonus plan tied to corporate profits. More companies are adopting variable pay plans tied to corporate profits, as well as individual performance. As this practice becomes increasingly common, it will become more difficult to compare both job offers and your future earning potential at different firms.

Consider not only the salary offer, but also the total compensation package. If you remain dissatisfied after this careful consideration, make a reasonable counter-offer based on your abilities and accomplishments. Also, consider asking for improvements in other aspects of the compensation package. For example, ask the employer to compensate for a low salary with a sign-on bonus. However, a sign-on bonus is a one-time event. Your starting salary will affect your compensation for the rest of your career with the firm.

Even if the salary offer is more than satisfactory, experienced chemical professionals may want to ask for a sign-on bonus. One circumstance in which you should consider this is when you are when moving from a moderate to a high cost-of-living area. Another is when you know your current home will be on the market for a long time. Yet another is when the company's funding of relocation costs is limited.

Unfortunately, employers seldom offer means of compensation such as sign-on bonuses to chemical technicians. Personal-performance bonuses are either part of every employee's variable pay package and tied to corporate business performance or are reserved for high-level executives.

If it could affect you, verify that your employer will pay you for any periods of active duty in the military. Usually this compensation is on the basis of the employer making up the difference between your regular salary and your military pay.

Retirement and Health Plans

These plans are not negotiable because federal law prohibits discrimination in favor of individuals or specific groups of employees. However, many employment plans have cafeteria-style options. Be sure you understand these and determine which would be best for you. The human resources department can advise you and answer questions. Consult them before accepting the job offer.

Relocation

Relocation can involve considerable sums of money and serious emotional issues. Beside basic moving costs, you may have some unusual ones you may wish the company to fund. These can include unusual transportation costs such as the fees to move a boat or a horse.

Your local housing market, your spouse's employment situation, or family members' education may delay your family's relocation. Therefore, you may need to make a down payment on a new house and cover monthly mortgage payments on two homes before you can sell your previous residence. Even if you rent an inexpensive apartment while your family lives in your old home or apartment, your living expenses have substantially increased. Negotiate with the employer to cover these additional living costs.

Some firms will have your old home assessed and purchase it at this assessed value. This can be very convenient; you avoid the stress of selling your old home while relocating and trying to make a good start on your new job. You will have the down payment you need for a new home and it may be easier for your family to move with you to the new location.

You can negotiate other possible means of obtaining a new home down payment. These include an interest-free loan from the new employer for a reasonable period (six months or so). The amount of the loan would be based on the value of your old home and the balance due on your mortgage. The term would be based on a reasonable estimate of the time needed to sell your current home. Negotiating a sum to cover simultaneous housing costs in two cities is difficult. Firms often reserve this benefit for high-level employees. However, persuading the company to pay travel costs for a reasonable number of trips to visit family members before they relocate is an option.

You may find you cannot move into your new home as soon as you desire. Many companies will cover hotel expenses, usually for not more than one month, in this situation.

Outplacement assistance for a working spouse is another important concern. Determine if the employer will provide any assistance to help your spouse obtain an equivalent position in your future area of residence. Maintaining two incomes can be critical to meeting your family's financial obligations and making all family members satisfied with the relocation. This is another reason to make this issue part of your negotiation. Also, long-distance marriages can cause great strains on relationships. Therefore consider this option very carefully if both you and your spouse are pursuing careers before committing to a relocation.

Professional and Education Costs

Many chemical professionals belong to technical societies such as the American Chemical Society and the American Institute of Chemical Engineers. Many are also active in industry-specific groups such as the Chemical Specialties Manufacturers Association, the Society of Petroleum Engineers, and the Technical Association of the Pulp and Paper Industry. Some professionals will want their employer to pay membership fees and meeting attendance costs associated with active participation in these groups. Corporate policies vary on covering these membership costs, but are sometimes negotiable.

Continuing education to stay up-to-date in your profession and broaden your skills is becoming ever more necessary. You may wish to obtain a commitment from the employer to pay these expenses.

Senior executives often negotiate severance agreements before they are even employed. This involves specifying a severance payment proportional to your salary for a specified period. Unless your job level is one at which such agreements are common, do not raise this issue.

Vacation Time

This issue should not be raised until after you have received a job offer. Formerly, no matter what your previous job history, new employees started with a standard vacation period, usually two weeks. This has become negotiable for experienced job offer recipients receiving more than two weeks vacation from their current employer. You may be able to negotiate vacation time equal to that you received in your former job. However, do not expect to receive more.

Unemployed candidates receiving job offers will find themselves in a weak position to negotiate additional vacation time. This is because, while salaries and other job-offer factors can remain confidential, vacation time seldom does. Current employees receiving less vacation time than a new hire can resent this. This can weaken the spirit of teamwork on the job. However, if vacation time is an important personal issue, the job offer recipient should raise the issue. An employer's refusal to negotiate will not leave you any worse off.

Starting Date

Having negotiated to improve your job offer, you now need to set a starting date should you accept the position. Even if you are currently unemployed, this requires serious consideration. If you are relocating, you need to allow yourself enough time to deal with the many details associated with moving. These include packing, closing the deal on your new home, accepting delivery and unpacking your possessions, and getting your telephone and utilities connected. If you have to take off work to attend to these details, you may give your new supervisor and coworkers a poor impression of your organizational skills and commitment to your new job.

Other Factors

There are additional factors that your employer may have little control over but will be important to your future job satisfaction and personal happiness. These factors are quality-of-life issues that relate to the suitability of the new location as a place for you and your family to live. This list includes the local cost of living, quality of area schools, cultural amenities, athletic participation opportunities and many other concerns that will vary from individual to individual. You can determine many of these factors if the employer offers you a second trip to familiarize yourself with the area. Many employers, particularly large companies, will do this after they have made you a job offer.

Counteroffers

There is one exception to job-offer negotiations. Suppose you are not completely satisfied with your job offer. In this situation, you may be tempted to inform your current employer of your dissatisfaction in the hope of receiving a counteroffer. Resist this temptation. However, if you are an outstanding performer, you may receive an unsolicited counteroffer when you give notice. Again, resist the temptation to accept. The manager may try to make you feel disloyal or tell you about future wonderful rewards. Counteroffers can increase your salary and improve other aspects of your job. Do not succumb to the temptation. Instead, ask yourself why the manager is only now offering to improve your situation.

Be aware of the manager's motives in making you a counteroffer. Sure, the managers regret the loss of a valuable employee—but more for their own sake than yours. Your departure will inconvenience your employer and delay completion of your projects. The employer may wish to keep you just long enough for you to complete an important project or for them to hire a suitable replacement. At the very least, your company loyalty will be suspect in the future. This can reduce your job security if you stay and your company later decides to reduce staff.

Your safest response to a counteroffer is to thank the employer for the counteroffer and say that you are committed to the new job and cannot back out. Regard an unsolicited counteroffer as a compliment, but stick to your original plan to leave. If you have doubts, consider the reasons you decided to find another job in the first place. Does the counteroffer really out-weigh these reasons?

The Last Details

At this stage you have decided to accept the job offer and are eagerly looking forward to changing jobs. However, before accepting the job and giving notice to your current employer, review your employment agreement. It is important to understand what you cannot do when you change jobs. You may find that the agreement precludes you from working for competitors in certain technology areas or places restrictions on what you can work on in a new job. These "noncompete" agreements are enforceable. While they affect relatively few job changes, if they affect you, take them very seriously.

Before accepting the job offer and giving notice, discuss with your future supervisor any restrictions your employment agreement may place on your responsibilities in your new job. If they are serious, you may find the employment offer regretfully withdrawn. Do not regard this situation as a total loss. You have not endangered your current job. Your professional attitude and the fact that the new employer was impressed enough to make you a job offer may lead to future employment opportunities at this firm.

The Exit Interview

You have now accepted a job offer and given notice to your current employer. Now you will have to endure one or more exit interviews. These are uncomfortable situations for all involved. In many cases, you will feel at least some regret as you reflect on the good aspects of your job, the occasions you enjoyed, and the coworkers you like. However, as noted in the discussion above on counteroffers, never let your decision to leave be changed by matters discussed in your exit interview.

Expect the employer to ask you why you are leaving. Prepare an answer to this question. Focus on the challenges and opportunities of your new job. Make only the briefest and most emotionally neutral comments about why you are dissatisfied with your current job. Do not volunteer these comments; make them only in response to questions. The interviewer may be sincere in wanting to know why you are leaving so they can correct the situation as it applies to other employees. However, they may take negative remarks personally or pass them on to others who will. You can expect resentment when you give notice but you do not want to create enemies. You may see former supervisors and coworkers at professional society and trade association meetings. You may work with them in activities of these organizations. You may even want the opportunity to work in another job for this employer later in your career.

For all these reasons, it is important to leave ethically with style and class.

How can you do this? Begin by making sure your work is up-to-date and all necessary reports are written. Give your employer adequate notice; two weeks is customary. However, be aware that some employers will demand that you leave the premises immediately. Be aware if your employer has a history of doing this and be prepared for this eventuality.

Inform your firm's patent department of your departure if you have any pending patent applications that need to be filed. It will be much easier to prepare the applications and get the necessary documents signed if you are still readily available. When you give notice, offer to train others in fulfilling your responsibilities. Take care of the details such as returning your copies of confidential reports and lab notebooks and returning all your company library books. Say goodbye to coworkers when you do this.

Avoid the temptation to tell coworkers why you are really leaving. Do not say anything to contradict what you said in your exit interview. Regard your exit interview as a confidential discussion and do not discuss it with coworkers. Your complaints can leave you with the label of a "bad apple" that the employer is better off without. Instead, when asked why you are leaving, focus on the new opportunity while noting how much you valued the questioner's work contributions and friendship. Remember, your immediate coworkers probably have become familiar with your workplace situation just by proximity. They usually know why you are leaving and often envy you for having the nerve to take the initiative into your own hands to make a job change. However tempting, do not urge any but close friends to change jobs too. (And if you should do this, do not have the conversation on company premises. Also, be sure you can trust your friend's discretion. Have this discussion because your friend is unhappy or frustrated in his career, not because you want to get back at your ex-employer.)

CAN YOU GO BACK AGAIN?

An increasing number of technical professionals are being recruited by their former employers. While most of these professionals are managers, marketing specialists, and sales professionals, some are researchers or applications and technical service specialists. Firms are finding that some people they hire have difficulties adapting to their corporate culture. Others are unknown quantities that do not perform to expectations. Both of these problems are less likely to occur if the company hires a former employee wellknown to people who are still with the firm. These former employees are being hired for both short- and long-term assignments.

What should you do if you receive an employment offer or a request to interview from a former employer? Enjoy the flattery of the offer but remind yourself why you left the firm in the first place. Was the firm slow to promote or otherwise reward performance? Was your supervisor difficult for you to work for? Did you basically like your job, supervisor, and coworkers, but leave for the offer of an increased salary or an exciting opportunity? Did you like your job but were discharged in a large-scale staff reduction? Determine if your old problems are likely to resurface should you rejoin the firm.

Determine the political aspects of rejoining your former employer. Will you be supervising people who were formerly in higher level positions than you? What interactions will you have with former coworkers who were difficult and decreased your job satisfaction? Do you understand the organizational changes and cultural changes that may have occurred in your absence? Will the fact you left the company affect your career advancement?

Talk to others who used to work for this company and have returned to determine what difficulties they encountered on returning to their former employer. Be sure to determine their assessment of their career prospects. In addition, all the factors discussed early in this chapter will apply. Consideration of all this information will help you to decide whether to return to your former employer.

Of course, the issues will differ if you are returning to your former employer for a short time only, a year or less. In this case, factors that could cause you to reject the offer to return to a former employer may have less force compared to the opportunity to earn some additional income and the challenge of the job assignment itself.

While you can return repeatedly for short-term assignments, do not expect to resign a long-term job a second time and be recruited again by your former employer.

FINAL COMMENTS

Evaluating job offers, negotiating better terms, and deciding whether to accept a job offer requires much time and effort. However, the results will affect the rest of your career—both in financial and professional terms. Be decisive, but take the time you need to make these important decisions. Consult your spouse and family members who will be affected by your decision. Finally, when you make your decision, celebrate it. Beginning a new job certainly deserves a celebration dinner or party. Avoid gloating, but include valued mentors and colleagues in your celebration.

Additional Reading

Chapman, J. *Negotiating Your Salary: How to Make $1,000 a Minute*. Ten Speed: Berkeley, CA, 1996.

Krannich, R. L. and Krannich, C. R. *Dynamite Salary Negotiations: Know What You're Worth and Get It!*. Impact: San Luis Obispo, CA, 1997.

Miller, L. E. *Get More Money on Your Next Job: 25 Proven Strategies for Getting More Money, Better Benefits, & Greater Job Security*. McGraw-Hill: New York, 1997.

Simon, M. B. *Negotiate Your Job Offer: A Step-by-Step Guide to a Win–Win Situation*. Wiley: New York, 1997.

PART IV

BACK
TO
THE
BEGINNING

18

CAREER ASSESSMENT AND PLANNING

Just as companies have business plans and annual goals, chemists should have a career plan and both long- and short-range goals. Your short-term goals are largely, but not completely, a job plan related to your current employment. Your long-term goals can include many objectives not connected to your current job such as a change of employment and retirement. Planning involves both goals and the actions you take to accomplish those goals. Career and job planning and assessment should be a continuous process. However, scheduling a time for career planning and assessment at a particular time of the year can add depth and discipline to the process. The approach of a new year and the tradition of New Year's resolutions makes December an ideal time to assess your professional accomplishments, learn from the past year's experiences, and plan for the following year. Another good time for this is shortly before your annual performance review. This process will improve your morale and self-image as you define your year's accomplishments. During your performance review with your supervisor, you will be able to discuss accomplishments and goals in a more organized way than if you make little preparation. Your career planning will also help you improve your productivity and aid your professional growth and career advancement.

REVIEW THE PAST YEAR

Begin by reviewing the past year. List your accomplishments. If you had goals for the year, review them to see how they compare to your accomplishments. Note the factors that made these accomplishments possible. This includes your personal and professional strengths and workplace factors. Workplace factors include abilities of your coworkers, your employer's facilities, and opportunities offered by your employer. Do not neglect professional society accomplishments. Do the same for your disappointments.

You may wish to review your self-assessment with your supervisor, mentors, highly trusted colleagues, and your family. See if their opinions of your strengths and weaknesses are similar to yours. If their opinions differ from yours, find out why. Some reassessment may be in order.

◈ Update Your Resume

> Your annual career review is an excellent time to review and update your job-hunting materials. Add your accomplishments to your resume. Add examples of your accomplishments and strengths to a file that you can draw on in writing cover letters and preparing for interviews.

By determining the factors responsible for your successes, you will identify strengths on which to build for a professionally successful and fulfilling new year. When you assemble this list, also consider what strengths and resources you did not exploit in the past year. For example, suppose you have strong presentation skills. If you made few oral presentations, you may not be capitalizing on this ability.

Also include the strengths of your coworkers and employers in your assessment. As chemists, engineers, and technicians work increasingly in teams, each person's individual success becomes tied to that of the entire team. Highly capable team members will make your own job easier and your goals more attainable. Just as you want to capitalize on your own strengths, you want to capitalize on the capabilities of your coworkers. For instance, suppose as a chemical salesperson, you have a coworker with advanced computer skills. Did you capitalize on her abilities in mastering new computer sales reporting programs mandated by your employer?

Also include the organizational resources of your employer in this list of strengths. Such resources include colleagues and advanced instrumentation in other departments and company consultants. Does your company fund university research relevant to your work? If so, are you actively involved in the program providing input and quickly learning about research developments?

Identify Areas for Improvement

In addition to the strengths and resources that you can build upon, identify the abilities you need to strengthen and the weaknesses you need to overcome. Your list of your disappointments will help you in this. Unachieved goals for past years may be due to poor planning and organization, inefficient use of time, insufficient coordination with coworkers, and other factors. Poor interpersonal skills may handicap you in accessing abilities and skills of your coworkers. Some causes of your unachieved goals may be external. These include insufficient abilities of coworkers and lack of corporate resources.

Identifying the factors responsible for your previous career disappointments may enable you to avoid them in the coming year. For instance, suppose you decide you have poor report writing or time-management skills. Identifying the problem can enable you to eliminate it by taking short courses and making a concentrated effort to improve. Should these disappointment factors be deep-seated, you may be able to plan your goals to reduce the effect these factors have on your career. There may be other ways to overcome them. For example, networking can enable you to find people with technical skills your coworkers lack. Outsourcing can enable you to overcome problems caused by a lack of corporate resources.

SET GOALS FOR THE COMING YEAR

By late in the year, you probably know your project assignments and coworkers for the coming year. Armed with your previous year's assessment, you can enter the new year with goals and a job plan that will let you capitalize on your strengths and rectify or overcome your weaknesses.

Setting goals will help you focus your work and professional activities to keep your career on track. Goals should be clearly defined. You should be able to measure them. For example, a sales person has the goal of increasing sales of product A to $25 million. He will know exactly when he meets his goal. Similarly, a bench chemist may have the goal of completing a manufacturing process to produce product A at a certain cost. Laboratory and scale-up results will tell him whether or not he has accomplished his goal.

Your goals should be challenging, but realistic. For instance, developing and commercializing one new catalyst in the coming year may be a realistic goal while developing three new ones is not. Most of your goals should contribute to your employer's goals: delivering improved products and services to customers while increasing profits. However, you should have additional goals that support your own professional and personal needs. These goals can include acquiring specific new or improved job-related skills. This may involve mastering a new area of chemical technology, learning how to use new computer software, or acquiring new job-related interpersonal skills. They may also include taking a certain vacation or accomplishing goals in your hobbies and recreational activities.

Plan to Improve Your Skills

Continuing education is critical to professional growth and is an essential part of any career plan. Many skills are transferable to other fields and different industries. These include laboratory skills, scientific knowledge, and organizational skills. In planning for the coming year, determine what is necessary to keep up to date in your specialty. Use a current awareness service to alert you to new papers in your field. However, conscientiously reading the journals may not be enough. You may wish to attend one or more technical conferences in your field. Include this in your career plan. Let your supervisor know you are attending these meetings as part of your career plan. Explain the advantages to both yourself and your employer in attending these conferences. The early notice and justification will greatly improve your chances of obtaining approval to attend.

As noted in chapter 5, professional meetings are excellent networking opportunities. Most professionals can benefit from making a conscious effort to improve their networking skills and expanding their network. Writing in *The Joy of Science*, Carl Sindermann defined networking as regular and frequent discussion with peers and colleagues. Today this discussion can take place by letter, fax, e-mail, or telephone. Company seminars and seminars at local universities, as well as regional and national technical conferences provide opportunities to maintain and expand your network.

Other skills, beside your specialty, are important to career success and fulfillment. Organizational skills let you function productively on the job. These skills include time-management, managerial, and teamwork skills, and the interpersonal skills that make you a well-liked and sought-after coworker. While staying up to date in your specialty is critical, it is also important to develop skills in these other areas. These more general skills can be critical to obtaining new employment or a sought-after reassignment.

Professional society and trade association activities offer opportunities to develop management, teamwork, and interpersonal skills. Thus, these activities should be an important part of your career plan. Many of these activities require multiyear commitments. Specific ways to participate include organizing a symposium, chairing a conference session, and participating in governance activities.

Consider what opportunities might be available in the coming year for cross-disciplinary or cross-departmental assignments. These can have many career benefits. You will broaden your technical and business expertise. More managers will notice your abilities and accomplishments. Also, these assignments are a way to add variety to your job.

Careers are no longer a smooth upward progression in a single department or company (see chapter 1). Consider opportunities for both upward advancement and lateral transfer in developing your career plan. Suppose a researcher knows that the business unit he serves is becoming a cash cow with few R&D opportunities. He would be wise to include exploring lateral transfer opportunities in his career plan.

Consider how you might better organize and manage your work assignments in the coming year. This assessment may identify a need for better time management. For example, a bench chemist should consider how to empower her technicians to perform some of her job assignments. These could include data and laboratory management and some report writing. The chemist can then have more time to focus on achieving other goals. These duties can enhance and expand technicians' job skills. This makes them more valuable both to the chemists they work with and to the company as a whole. Their broader range of skills can open up new opportunities for advancement and new career paths.

Review what opportunities there may be in the coming year to patent the results of your work and publish or present papers. Do not limit your publication only to highly technical journals. Consider publishing your work in trade association and industry magazines. Your published article may be less technical, but have more commercial impact. Include both trade association and professional society meetings in your assessment of presentation opportunities.

Balance Your Personal and Professional Lives

Your plan for the coming year can also help you coordinate your professional activities with your personal life and achieve a better balance between the two. For instance, by coordinating a business trip with vacation plans, you may be able to visit family or see an interesting part of the country. A Saturday night stay could even reduce your corporate travel expenses. Planning will enable you to budget time for civic, charitable, and other personal activities that are important to you. For example, your New Year's goals could include taking a certain vacation or accomplishing certain things in your hobbies and recreational activities. The focus that your plan gives you will allow you to be more productive at work and give you more time for your personal life.

One way to help balance your professional and personal lives is to delegate more assignments and tasks to others. Your annual review is a good time to consider which of your tasks could be delegated to others. This could free your time to focus on tasks only you can perform, make your work schedule less hectic thus freeing time for personal pursuits, and enrich the jobs of the individuals to whom you assign your former tasks.

LONG-TERM PLANS

The discussion to this point has related to short-term plans and is essentially a job plan. However, you have a career and a life, as well as a job. Beside setting short-term personal life goals as described above, you need to have a long-term plan to include career and personal goals. Long-term career goals can include objectives such as engaging in an alternative career, working for another company, getting a job in a specific part of the country, and so on. Personal aspects of your long-term plan may include buying a new home, financing your children's college education, relocation after retirement, vacation planning, and many other possibilities. These long-term goals may require that you include specific activities in your short-term planning. Career-related goals can include continuing education to master skills you will need in an alternative career. For example, someone planning to start a consulting business after retirement could take some basic business courses in accounting and marketing. Other short-term activities related to long-term goals include:

- developing a personal network with a view to job hunting later
- trying to get a transfer to a different department as a step in moving towards an alternative career
- presenting papers at conferences with a view to improving your credentials for later job hunting.

Short-term aspects of long-range personal plans could include many different goals. The most obvious one is establishing or continuing a savings plan.

EXECUTING YOUR PLAN

Using effective time management (chapter 2) and practicing your plan will enable you to translate your goals into accomplishments. Now that you have your plan, do not let it gather dust. Use it as a master blueprint in planning your weekly and monthly activities.

Organize your weekly work activities around accomplishing your goals. Frequently referring to your plan will help you keep your work and professional life focused.

Keep track of deadlines to help you accomplish your goals in a timely manner. Schedule backward from your deadline. Scheduling backward will give you an estimated starting date to begin work on your project. It will also let you know if your deadline is feasible. Do not use this estimated starting time as an excuse to delay the start of the project. Also, do not establish too tight a schedule. Try to allow for slippage that can arise due to the need to perform additional work, confirm results, fill in gaps in your data, instrument malfunctions, and other unforeseen circumstances. Setting intermediate goals to serve as milestones will help you keep on track. The time management section of chapter 2 will give you additional suggestions on how to schedule projects.

Many times a project such as presenting a paper at a conference can get repeatedly delayed in the rush of day-to-day work. The result is the paper never gets submitted or is withdrawn. Effective time management with a focus on achieving your goals can prevent this frustrating and demoralizing situation.

Changes in your employer's priorities or your job assignment could require changing your plan. Should this happen, do not be overly rigid. Also, look for ways you can utilize the effort you put into your now outdated plan. For instance, if your employer terminated a project, interesting technical data may still merit being documented in an internal report. Should circumstances warrant, presentation of these results at a conference or publication in a journal may be feasible.

Unforeseen Circumstances

The best short-term plan can become irrelevant if you become an unexpected victim of a corporate downsizing or takeover. It probably is not necessary to develop a detailed plan for this circumstance. Still, a flexible long-range career plan plus financial planning can help you be prepared should unexpected job loss occur. Your long-range plan will aid you in developing a course of action should you lose your job or face unexpected reassignment. Financial planning for unexpected job loss is critical, as is keeping your resume up to date. It is often difficult to focus and energize yourself for job hunting while you are still traumatized by your job loss. Having a resume that needs little updating helps greatly in beginning your job hunt. When you review your past year's accomplishments, update your resume also.

Contingency planning for retirement can help make an unexpected early retirement less traumatic. Major decisions such as selling a home and making a long distance move should not be made hastily or while you are emotionally upset.

Your short- and long-range plans will help you decide how to respond to unforeseen events in your professional and personal lives. Examples include whether to accept a job transfer to another department or city and your response to an offer to teach a night course at a local university. A very time-consuming activity such as a request to write a chapter for a technical review book may have little relation to your job goals. However, it may provide deep personal satisfaction. Writing it can help you accomplish a career goal such as improving your professional credentials for a job change or interfere with personal goals such as taking a long vacation to travel in Europe. Review your short- and long-term plans to determine if writing the book chapter will help you accomplish your goals. Thus, your career plan can aid you in deciding whether or not to make the time-consuming commitment to write the book chapter.

By following the advice "plan the work and work the plan," you can improve your career advancement, increase your technical accomplishments, and enjoy your work more.

Additional Reading

Covey, S. R., Merrill, A. R., and Merrill, R. R. *First Things First*. Simon & Schuster: New York, 1964.
Douglass, M. E. and Douglass, D. N. *Manage Your Time, Your Work, Yourself*. American Management Association: New York, 1993.

Drucker, P. F. *The Effective Executive*. Harper & Row: New York, 1966.

Greissman, B. E. *Time Tactics of Very Successful People*. McGraw-Hill: New York, 1994.

Moskowitz, R. *How to Organize Your Work and Your Life*. Doubleday: Garden City, NY, 1981.

Wilson, S. *The Organized Executive*. Warner: New York, 1964.

Index